Fungi: A Very Short Introduction

VERY SHORT INTRODUCTIONS are for anyone wanting a stimulating
and accessible way into a new subject. They are written by experts, and
have been translated into more than 40 different languages.

The Series began in 1995, and now covers a wide variety of topics in every
discipline. The VSI library now contains over 450 volumes—a Very Short
Introduction to everything from Psychology and Philosophy of Science
to American History and Relativity—and continues to grow in every
subject area.

Very Short Introductions available now:

Available soon:

GOETHE Ritchie Robertson
ENVIRONMENTAL POLITICS
    Andrew Dobson
MODERN DRAMA
    Kirsten E. Shepherd-Barr

THE MEXICAN REVOLUTION
    Alan Knight
HISTORY OF CHEMISTRY
    William H. Brock

For more information visit our website

www.oup.com/vsi/

Nicholas P. Money

FUNGI

A Very Short Introduction

OXFORD
UNIVERSITY PRESS

OXFORD
UNIVERSITY PRESS

Great Clarendon Street, Oxford, OX2 6DP,
United Kingdom

Oxford University Press is a department of the University of Oxford.
It furthers the University's objective of excellence in research, scholarship,
and education by publishing worldwide. Oxford is a registered trade mark of
Oxford University Press in the UK and in certain other countries

© Nicholas P. Money 2016

The moral rights of the author have been asserted

First edition published in 2016

Published in the United States of America by Oxford University Press
198 Madison Avenue, New York, NY 10016, United States of America

British Library Cataloguing in Publication Data
Data available

Library of Congress Control Number: 2015945428

ISBN 978-0-19-968878-4

Printed and bound by
CPI Group (UK) Ltd, Croydon, CR0 4YY

Contents

Acknowledgements

The author wishes to thank Anna Heran, Curator at the Lloyd Library and Museum, for her expertise in scanning original sources from the outstanding collection of mycological books and periodicals in Cincinnati. Additional work in preparing the figures for publication was performed by the author's research collaborators, Maribeth Hassett, Miami University, and Mark Fischer, Mount St. Joseph University. The manuscript was proofread by Allison Davis, who was very effective in identifying topics that deserved clearer elucidation.

List of illustrations

Fungi

Chapter 1
What is a fungus?

Defining the fungi

Fungi are peculiar organisms. They do not seem to move, their fruit bodies pop up overnight, and they have no visible means of feeding themselves. Early botanists considered the immobility of mushrooms as a sign of simple plant life, which explains why mycology—the study of fungi—is included in the traditional purview of botany. In contrast to plants, however, fungi do not have chlorophyll, lack leaves and roots, and never form flowers, fruits, and seeds. The combination of these non-animal and non-plant characteristics with the poisonous and hallucinatory nature of some mushrooms explains why fungi have been associated with witchcraft and supernatural beliefs. No other part of biology has encouraged this kind of folly. Putting aside the paranormal, this short book concentrates on the remarkable facts of fungal biology and the global significance of these enchanting organisms.

Until the 17th century, natural historians had no understanding of fungi beyond what little had been gleaned from looking at mushrooms and considering their edibility. This changed with the invention of the microscope and Robert Hooke published striking drawings of fungi sprouting from a 'mildewed' book cover and infecting the leaves of a rose bush in his famous *Micrographia* (1665). Hooke was uncertain about the identity of fungi, describing the anatomy of mushrooms in the section of his book on sponges. Later investigators were fascinated by the movement of fluid they observed inside fungal cells and wondered whether these lively filaments were produced by a strange group of animals. Linnaeus found the fungi very perplexing and classified them with worms in the 1767 edition of his *Systema Naturae*, but the long-standing consideration of fungi as primitive plants survived this zoological interlude.

The modern scientific definition of the fungi draws upon information from a variety of sources. Three principal characteristics unite the fungi: they are eukaryotes, which feed by absorption, and reproduce by forming spores. The term 'eukaryote' refers to cell structure and means that an organism's genetic information is housed inside a structure called a nucleus. Animals and plants are eukaryotes too, along with single-celled amoebae, microscopic algae, slime moulds, and seaweeds. Other forms of life—bacteria and archaea—are prokaryotes that do not have nuclei. Fungi feed by digesting materials produced by animals and plants. They do this by releasing enzymes that break down complex substances into smaller molecules including sugars, amino acids, and fatty acids, which are transported into the cell. This feeding mechanism is referred to as 'osmotrophy'. We perform a similar process when we use our digestive enzymes to break down carbohydrates, proteins, and fats in our digestive system. Spore formation is the third unifying characteristic of fungi. Fungi produce an awful lot of spores: a single mushroom can release 30,000 spores per second from its gills and millions of tons of these tiny particles are dispersed in the atmosphere every year.

Genetics and evolutionary origins

In addition to the trinitarian description of fungi as *eukaryotes that feed by absorption and reproduce by spore formation* we can separate them from other forms of life using genetic and cell biological criteria. Comparisons between key genes in different organisms—revealed in strings of the universal four-letter DNA alphabet (A, T, G, C)—provide a powerful measure of relatedness. The sequences of these genes in species of fungi that form the fly agaric mushroom (red cap with white spots, *Amanita muscaria*) and the death cap mushroom (greenish cap, *Amanita phalloides*) are more similar to each other than the sequence of either mushroom is to the DNA of baker's yeast (single-celled fungus, *Saccharomyces cerevisiae*). If we compare the genes of all three fungi with the sequences of the same genes in fish or humans, it is apparent that the mushrooms and yeast are more similar to one another than any of them is to an animal (Figure 1). This finding reflects the relatedness of the fungi and their separation, over the course of hundreds of millions of years, from the animals. This use of genetics to reveal evolutionary kinship between organisms is the science of molecular phylogeny.

In practice, it is quite difficult to study the evolutionary relatedness of organisms as distant as mushrooms and animals because genes that are useful for comparing different fungal species do not work as well for animals. Nevertheless, these big picture studies of evolution show, in fact, that fungi are more closely related to animals than they are to plants. According to the current view of biological diversity, fungi and animals occupy one of the seven or eight major branches of eukaryote life. Each of these branches is given the name of a taxonomic supergroup and the fungi and animals belong to the Opisthokonta supergroup.

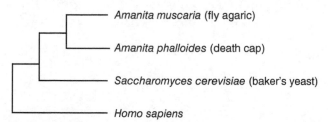

1. **Evolutionary relationships between fungi and animals illustrated in a horizontal 'tree'.**

The name opisthokont refers to the formation of cells that swim using a single posterior tail or cilium. Human sperm cells move in this fashion and aquatic fungi called chytrids use the same propulsive mechanism. Most fungi do not produce cilia, but their presence in chytrids, which are viewed as relatives of the earliest fungi, is seen as strong evidence of the common ancestry of fungi and animals (Figure 2). Biologists have designated Kingdom Fungi and Kingdom Animalia as distinct groups within the supergrouping of the Opisthokonta.

The emergence of the fungi as a distinctive group of organisms is estimated to have happened between 760 million years ago and one billion years ago. This fuzzy Precambrian origin, somewhere in the early Neoproterozoic Era, is inferred from genetic differences between living fungi. Estimates of the rate of change of genetic sequences serve as a 'molecular clock' for calculating the time when particular groups of organisms diverged from one another. Fossils of fungi are helpful for calibrating these clocks. These include tiny mushrooms preserved in 100 million-year-old

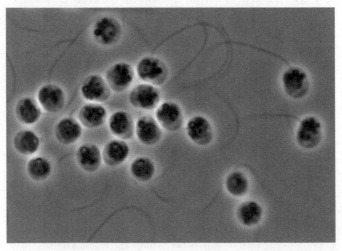

2. Zoospores of a chytrid fungus. Each spore has a single flagellum.

Cretaceous amber, colonies of fungi related to chytrids in 400 million-year-old Devonian chert, and large spores in 460 million-year-old rocks deposited in the Ordovician.

The Cretaceous mushrooms are beautifully preserved, but their resemblance to fruit bodies that we find in the woods today shows that this group of fungi evolved much earlier. This conclusion is supported by the discovery of 330 million-year-old fossils of the type of cells that that are characteristic of mushroom colonies. The chytrid relatives are more interesting because they show that fungi were flourishing alongside diverse plant communities that spread across the Devonian landscape. The shape and size of the Ordovician spores suggests that they were produced by a species belonging to a group of fungi that form associations with plants called arbuscular mycorrhizas. This suggests that early fungi were engaged in supportive partnerships with the first land plants. Together, these precious fossils demonstrate that fungi and plants have occupied the same habitats for almost half a billion years.

The fungal cell

The Precambrian separation of the fungi and the animals was probably driven by their pursuit of different ecological opportunities and this evolutionary divergence led to the development of features of cell biology and physiology found only in fungi. These exclusive characteristics are layered upon the essentials of eukaryote structure and function that include the expression of genes contained in a nucleus and power generation by organelles called mitochondria.

The fluidity of fungal membranes is maintained by a lipid molecule called ergosterol. Cholesterol does the same thing in animal membranes and the absence of ergosterol in humans makes it a good target for drugs used to treat fungal infections (Chapter 7). The lipid membrane of fungi is surrounded by a cell wall whose chemical composition distinguishes fungi from plants and other organisms with cell walls. Plant cell walls contain cellulose. Fungal walls contain chitin, chains of sugar molecules called glucans, and mixtures of proteins. Chitin, which forms the exoskeleton of insects and is widespread in other animal groups, is made from chains of two kinds of modified sugars called amino sugars. The chitin content of the fungal wall can be quite low compared with the glucans, but chitin forms long strands that have tremendous tensile strength. The wall needs to be strong because the cell contents are inflated to the same level of pressure as bicycle tyres—albeit with compressed fluid rather than air—and rupture if the wall becomes weakened. This 'turgor pressure' is an important characteristic of fungi and its role in growth is described later in this chapter.

Like other eukaryotes, the inside of the fungal cell (the cytoplasm) contains compartments surrounded by membranes that form a series of globules that transfer materials between the cell and the environment. This endomembrane system includes the endoplasmic reticulum, Golgi apparatus, elongated vacuoles, and small spherical vesicles. One of the main functions of the endomembrane system is to deliver new membrane and cell wall components to the growing cell surface. The growth process is also dependent on the arrangement of an internal framework of protein strands. This cytoskeleton is constructed from microfilaments of the protein actin and hollow microtubules of tubulin. Microfilaments and microtubules serve as tracks for molecular motors (natural nanomachines) that ratchet along their surface carrying endomembrane vesicles around the cell.

This essential set of structures is found in two kinds of fungal cells: filamentous hyphae, which grow by elongation and branching, and yeasts that multiply by forming buds (Figure 3). The cytoskeleton inside a hypha is organized lengthwise such that vesicles delivering new cell materials from the endomembrane system are transported towards the growing tip of the cell. When

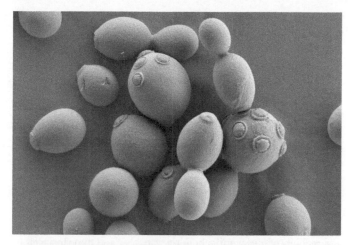

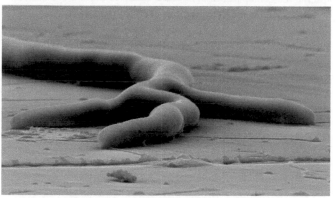

3. Scanning electron microscope images of (top) yeast cells of *Saccharomyces cerevisiae*, and (bottom) filamentous hyphae of *Aspergillus niger*.

the vesicles reach the end of the cell they fuse with the membrane, allowing it to expand. The turgor pressure within the cell tends to smooth the expanding membrane and extend the cell wall. This pressure results from the absorption of water by the cytoplasm. Growth is not simply a matter of pressurized inflation, however,

7

because there is a complex interplay between the level of turgor in the cell and the mechanical properties of the cell wall. The hypha controls the fluidity of its wall with great finesse, loosening the chemical bonds between the chitin and glucan molecules, allowing them to slip past one another without triggering the explosion of the cell.

The beautiful coordination of these cell mechanisms is apparent to anyone watching a hypha grow under a microscope. At high magnification, the tip of a hypha appears to extend at a constant speed while vesicles and other structures whirl around inside the cell. Some sense can be made of this confusing picture using fluorescent dyes linked to antibodies that attach to specific cell proteins. These dyes allow investigators to study the distribution and mobility of nuclei, mitochondria, vacuoles, and the cytoskeleton. One hyphal structure that has been studied for many decades is a cluster of vesicles at the tip of growing cells called the Spitzenkörper, or dark body (Figure 4). The position of this organelle corresponds

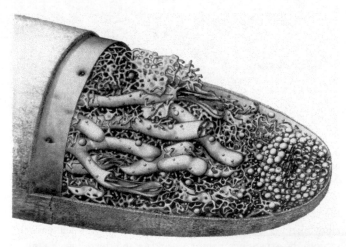

4. Diagram showing the interior of a hyphal tip packed with vesicles that form a structure called the Spitzenkörper whose position governs the direction of growth.

to the direction of hyphal extension and experiments suggest that its movement controls growth.

As hyphae elongate they form branches, creating an interconnected web of cells that expands at its edge (Figure 5). This is the fungal colony or 'mycelium', whose circular shape is manifested in the fairy rings of mushrooms and in the disc shape of skin infections caused by ringworm fungi. The expanding mycelium digests materials in its path and uses turgor pressure to push aside obstacles and penetrate solid food. When hyphae encounter sand

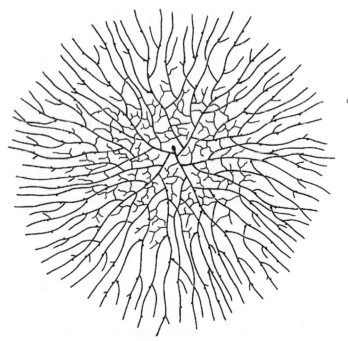

5. Young colony or mycelium of a fungus that has grown outwards from a single spore at the centre of the drawing.

grains and other hard objects they grow around them and produce branches to explore alternative routes. Rather than sensing traces of chemicals diffusing from potential food materials, mycelia spread outwards in all directions until they strike digestible objects. When this happens, the colony reacts by redirecting growth towards these locations.

Hyphae form a continuum of filaments throughout the colony. This interconnected structure allows a mycelium to be highly responsive to the distribution of food in its environment. There is, however, a potential weakness in this organization. Without an efficient repair mechanism, damage to the mycelium in one place—caused by a browsing snail, for example—would affect the whole colony through the rapid loss of turgor pressure as cytoplasm oozed into the soil. This apparent architectural flaw has been countered by the evolution of an elegant substructure to the mycelium. Most fungi produce partitions or septa along their hyphae that divide each cylinder into compartments. Each septum has a closable hole in the middle allowing cytoplasm to flow from one compartment to the next. In one group of fungi, called basidiomycetes, the septa are surrounded with caps formed from perforated membranes. These regulate the passage of organelles along the hyphae, acting as sieves through which larger structures are excluded. And, when damage occurs to one part of the colony, septa on either side of the wound are sealed by the membrane caps to localize 'bleeding'.

The succession of compartments that forms the mycelium is regarded as a multicellular structure. Individual hyphal compartments of some fungi contain single nuclei, but other fungi have multiple nuclei in each compartment. This version of multicellularity is unique to the fungi. Some fungi do not form septa along their growing hyphae and develop, instead, as a fluid-filled continuum of cylindrical cells containing multiple nuclei. In these fungi the entire colony is regarded as a single multinucleate cell, also known as a coenocyte.

The second common growth form of fungi is the yeast. This alternative cell type has a rounded shape and reproduces by forming buds on its surface. Hundreds of species of fungi grow in this fashion and are referred to as yeasts. The noun 'yeast' is also used in a narrower sense as the common name for a single fungal species, *Saccharomyces cerevisiae*, which is used in baking, brewing, and winemaking. Yeasts tend to flourish in fluids rich in nutrients, but they also grow on wet surfaces where they form gooey blobs of multiple cells. Most fungi are aerobes, meaning that they require oxygen to support the respiratory reactions that release energy from their food. Oxygen can become limiting for yeasts submerged in fluid and this stimulates fermentation reactions that release energy from nutrients under anaerobic (oxygen free) conditions. Fermentation is the process that generates alcohol in brewing and winemaking. Many of the fungi that grow on human skin and are found in our guts are yeasts. *Candida albicans* is a yeast that normally grows in the mouth, vaginal tract, and colon without causing any illness. It causes oral thrush in babies and vaginal irritation in adults when antibiotics upset the normal balance of microorganisms. *Candida* can also produce serious infections when the immune system is weakened.

A typical yeast cell is 0.003 to 0.004 millimetres (3 to 4 μm) in diameter, which is three times wider than a bacterium and half the size of a red blood cell. Fungal hyphae vary a lot in width, but most are around the same width as a yeast cell. Hyphae are visible as white threads when they grow in bundles (underneath a plant pot is a good place to look), but a microscope is needed to see an individual hypha or a yeast. This is the reason that we refer to fungi as microorganisms, even though their fruit bodies can be huge. Popular references to fungi as the largest and oldest organisms are based on the estimated size of mycelia of *Armillaria solidipes*, a 'honey fungus' that grows in forest soil and attacks conifers. The current champion has spread over ten square kilometres in Oregon in the last 2,400 years and may weigh as much as 35,000 tons. The largest umbrella shaped mushrooms have a cap diameter of one metre and are produced by colonies of fungi cultivated by termites in Africa and leaf-cutter ants in Central America. Most fungi are microscopic throughout their lives, forming tiny colonies that become dusted with spores.

The inconspicuous nature of most fungi is one of the reasons that we know so little about them compared with animals and plants. Most of the classical work in mycology concerned mushrooms and there was little attention to the development of the colonies that supported the fruit bodies. The introduction of culture techniques in the 19th century placed the science of mycology on a stronger scientific footing and a great deal was learned from experiments on a small number of species that could be grown easily in the laboratory. Studies on the fine structure of fungal cells using electron microscopy produced a wealth of information on growth mechanisms and cell division in the 1970s, and this work was extended through molecular genetic experimentation beginning in the 1980s.

The emerging molecular techniques were also applied to the study of phylogeny at this time and today's fungal biology is informed by the increasingly sophisticated methods for sequencing and characterizing genes. Using modern 'metagenomic' methods, for example, investigators are extracting and characterizing genes from the environment without growing fungi in culture. A new group of fungi, called the Cryptomycota, was discovered with these methods in 2011. The genes of these microorganisms have been identified in pond water, soils, marine sediments, and even in chlorinated drinking water. They appear to embrace more genetic diversity than all of the groups of fungi known previously, but almost nothing is known about their biology beyond the fact that they form motile zoospores.

The diversity evident in today's fungi has arisen through processes of evolutionary change across the hundreds of billions of generations of fungi that have lived and died since the Precambrian.

Sexual reproduction is widespread in the fungi but unknown in some groups. Natural cloning of fungi occurs when a nucleus divides by the process of mitosis and copies of the nucleus are packaged into spores. These asexual spores formed by mitosis are called conidia. Mutations in the genes carried inside conidia are a major source of the variation with which evolution operates and asexual or 'parasexual' mechanisms of genetic recombination produce additional novelties. Sexual reproduction occurs when mycelia of two strains of a fungus fuse with each other. Many fungi reproduce asexually and sexually, developing two or more kinds of spore depending upon the method of reproduction.

Fungal ecology

Fungi engage in all manner of close biological associations with other organisms. The scientific term for these interactions is symbiosis and they include relationships that appear to benefit both contributors (mutualism) and interactions in which one participant benefits at the expense of the other (parasitism). Fungal mutualisms include a range of physical connections with plant roots called mycorrhizas, including the arbuscular type preserved in the Ordovician fossil record. Lichens are another kind of fungal mutualism in which the cells of a photosynthetic microorganism— either a cyanobacterium or a single-celled eukaryotic green alga—are embedded in a lattice of fungal hyphae. Thousands of species of fungi form lichens in collaboration with a small number of photosynthetic partners. Fungal mutualisms with animals include symbioses that form inside the guts of invertebrates and vertebrates. Relatively little is known about these interactions, but the fungi participate in the breakdown of food within the digestive system of their hosts. Bucking the aerobic nature of most fungi, species of strictly anaerobic fungi live in the guts of herbivores where they digest plant fibre. Other kinds of fungi live in the human gut where they are thought to interact with resident bacteria and archaea in the metabolism of sugars.

Fungi are the most important cause of plant disease and are responsible for billions of dollars of crop losses every year. In some of these infections, the fungus feeds on living tissues without killing the plant. Other fungi begin by killing plant cells and feed on their dead contents. In a third disease category, the fungus employs both strategies by minimizing damage to the living plant in the early phases of infection and switching to more destructive behaviour later. All of these fungi can be described as plant parasites or as pathogens. There is some disagreement about the difference between parasites or pathogens, but these terms are usually interchangeable when we refer to fungi that cause disease. Epidemic fungal diseases of trees and staple crops have affected us throughout history, but even the least conspicuous plant is plagued by pathogenic fungi.

The Irish potato famine of the 1840s is often attributed to the spread of the potato blight 'fungus' *Phytophthora infestans*. This pathogen is not a fungus; it is a species of water mould more closely related to giant kelps and other brown algae than it is to mushrooms and yeasts. Nevertheless, water moulds have been studied by mycologists because they form hyphae that resemble the cells of fungal mycelia. Indeed, water moulds meet the functional definition of fungi as *eukaryotes that feed by absorption and reproduce by spore formation.* Biologists separate them from fungi on the basis of numerous differences in cell structure and genetics, which show that their resemblance is an example of evolutionary convergence, comparable to the formation of wings by birds and bats.

Fungi decompose plant tissues in forests and grasslands. They produce enzymes that catalyse the breakdown of lignin polymers that strengthen wood. Other fungi decompose plant tissues in rivers and lakes and produce spores beneath the water that attach to submerged leaves. Breakdown of partly digested plant tissues in herbivore dung is another fungal activity that is a critical process in the carbon cycle. Dispersal of these coprophilous fungi once the nutrients in the dung are exhausted requires the escape of their spores, passage through an animal, and deposition in a fresh deposit of faeces. Evolutionary solutions to this considerable challenge have produced gorgeous contraptions for discharging spores, including the squirt gun of *Pilobolus* that launches a capsule filled with 90,000 spores at a speed of 32 kilometres per hour over a distance of 2.5 metres (Figure 6). Scaled to human dimensions, this is equivalent to a nine-kilometre flight!

Some fungal pathogens of plants and wood decomposers may be exclusive vegetarians, but most fungi are very effective at breaking down animal proteins when they are fed them in the laboratory. This omnivory is reflected in the observation that a great variety of fungi are capable of infecting the tissues of animals with weakened immune systems. Infections of this kind are called opportunistic and, in rare cases, even involve fungi that form mushrooms. More prevalent fungal infections are also opportunistic, but they are established by species that show particular adaptations to combating the remaining host defences and colonizing our tissues. Human interactions with fungi can be harmful in other ways including poisonings by highly toxic mushrooms, exposure to 'mycotoxins' produced by fungi that cause food spoilage, and allergies stimulated by the inhalation of airborne spores. Human physiology is also affected by psychoactive compounds produced by 'magic mushroom' species that are sought by devotees of hallucinatory experiences. Drugs purified from these fungi have also been used in some fascinating neuroscience experiments that have illuminated brain function.

The yeast *Saccharomyces cerevisiae* has been used to make bread, ferment wine, and brew beer for millennia and fundamental practices of mushroom growing have changed very little for centuries. Traditional methods of working with fungi have been refined with the use of carefully controlled growth conditions and strains of fungi with specific properties. Genetic engineering has become part of the biotechnology industry in

6. The stalked sporangium of the dung fungus *Pilobolus kleinii*. The translucent stalk is a few millimetres tall and filled with pressurized fluid. The swelling at the top serves as a lens that focuses sunlight on pigments that allow the fungus to point towards the sun. Discharge of the black sporangium occurs when it breaks from the stalk and is blasted up to 2.5 metres through the air at a speed of 32 kilometres per hour.

which yeasts and filamentous fungi are used to produce antibiotics and other pharmaceutical agents, industrial enzymes, organic acids, and vitamins.

At a time when fewer and fewer biologists specialize in the broad study of groups of organisms, biologists who call themselves mycologists have become as scarce as professional entomologists and ornithologists. But fungi are studied by many scientists, including cell and molecular biologists who use fungi as 'model systems' for exploring fundamental questions about the way cells work. Indeed, fungi have been used in some of the most important experiments in modern biology including research that led to the 'one gene-one enzyme hypothesis' in the 1940s, work on the cell cycle in the 1970s and 1980s, and the sequencing of whole genomes in the 1990s. Other scientists who study fungi include plant pathologists who are just as interested in plants as the fungi that attack them, specialists in indoor air quality concerned with the allergens carried by spores, and palynologists who use historical deposits of fungal spores (along with pollen) as indicators of ecosystem and climate change.

Beyond the academic study of fungi, many people enjoy mushroom hunting and love to cook and eat bronze-capped ceps (porcini), golden chanterelles, honeycomb-headed morels, and other delicious edibles. The yin-yang nature of the mushroom, with the shared pulchritude of the poisonous and the edible, places a premium upon the art and science of mushroom identification. This makes an illustrated guidebook on fungi a viewing pleasure and a potential lifesaver. And with so many mushrooms, there is a great deal to learn. We turn in Chapter 2 to the wider diversity of the fungi as we consider the characteristics of the major groups recognized by biologists.

Chapter 2
Fungal diversity

The range of fungal complexity

Fungi are more diverse in their structure and behaviour than anyone would imagine from their representation in popular culture. Most people recognize that mushrooms come in a range of sizes and colours, that yeasts are fungi used in baking and brewing, and that other microscopic fungi cause vaginal irritation and athlete's foot. This snapshot of the fungi does not come close to representing the variety of the mycological world.

Consideration of three species that span the range of structural complexity among the fungi may be helpful in expanding our survey of this group of organisms. A fungus called *Olpidium brassicae* is about as simple as a fungus gets. *Olpidium* is a parasite of cabbage that uses swimming zoospores to infect the root cells of its host (Figure 7). Once inside a root, the fungus grows as a rounded pellet, or thallus, that feeds on its surroundings. Each thallus expels a new generation of zoospores after the nutrients in the infected cell are exhausted. *Olpidium* uses an intricate mechanism to penetrate the cells of its host but its structure is very plain compared with other fungi. The only deviation from the cycle of swimming zoospores, plant infection, formation of feeding thalli, and return to swimming zoospores, is

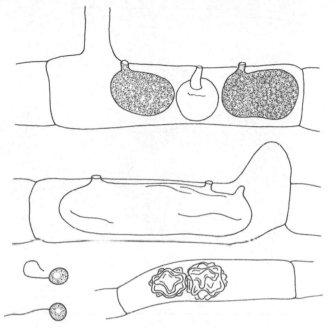

7. *Olpidium brassicae*, an aquatic fungus that infects cabbage roots using zoospores.

the formation of angular resting spores that can maintain the fungal population at the end of the growing season.

Spirodactylon aureum is a representative of the middle of the range of structural complexity (Figure 8). This fungus grows on rodent dung and forms spores within coiled stalks that project into the air. Each stalk is 1–2 millimetres in height and supports numerous coils. The whole structure gyrates when it is disturbed and looks like a miniature crystal chandelier. It has been suggested that the spores of this fungus are spread on the hairs of rodents that brush past the coils. Although spore formation in this fungus involves the development of this spectacular

8. *Spirodactylon aureum*, a zygomycete fungus that grows on rodent dung and forms its spores on stalks with a spectacular coiled structure. The fungus is shown in a variety of magnifications in this illustration. The spores are formed within the coils and are exposed in the drawing at bottom left.

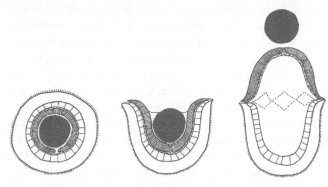

9. The artillery fungus, *Sphaerobolus stellatus*, sliced through the centre to show the multiple tissue layers. (Left) unopened fruit body; (middle) open fruit body with black capsule bathed in fluid within cup; and (right) capsule jettisoned from triggered cup.

apparatus, *Spirodactylon* does not create any kind of larger fruit body comparable to a mushroom for housing the cells that produce its spores.

The most complex species of the trio is *Sphaerobolus stellatus*, which has the common name of artillery fungus (Figure 9). Its closest relatives produce earth-stars and phallic mushrooms which will be described shortly. Colonies of *Sphaerobolus* feed on herbivore dung, mulch, and wet wood chips, and make spherical fruit bodies with a diameter of two millimetres. The anatomy of these little white beads is exceedingly complex. The centre of the fruit body is occupied by a capsule that contains ten million spores. A jacket made from six layers of interwoven hyphae surrounds the capsule and the whole structure looks like a tiny Scotch egg. When the fruit body is mature, the outer layers of the jacket peel open, separate into two cups, and hold the exposed capsule in the centre. Swelling of cells in the inner cup produces a compressive stress which is relieved when this elastic membrane flips outward with an audible 'pop', propelling the capsule into the air over a distance of up to six metres.

The largest structures produced by *Olpidium* are its thalli inside the cabbage roots. These can be seen with the low-power objective lens of a microscope providing a 100-fold (100×) magnification. The stalks of *Spirodactylon* are visible with the naked eye, but none of their intricate structure can be seen without a microscope. The fruit bodies of the artillery fungus can be spotted on the ground, and the magnification provided by a 10× hand lens (magnifying loupe) is sufficient to see the glistening capsule bathed in the fluid of the open cup. A microscope is needed to see the multilayered structure of the jacket and the spores inside the capsule. Light microscopes with magnifications between 100× and 1,000× are essential for studies on fungi and can be used to look at living cells. Electron microscopes provide considerably higher magnifications of dead specimens that are placed in a vacuum and bombarded with electrons.

Classifying fungi

Until quite recently, the visible characteristics of the fungi served as the only guide to identifying species and for sorting them into different groups. Long before scientists became interested in developing a formal system of fungal classification, illustrated studies of mushrooms were published with the aim of separating poisonous and edible species. Carolus Clusius authored the first detailed study of mushrooms in 1601. His work was extended by a Flemish priest, Franciscus van Sterbeeck, in a very influential book, titled *Theatrum Fungorum*, published in 1675. Later scholars separated mushrooms by differences in shape, size, and colour and the first schemes for classification assigned these putative species into groups with gills, pores, or teeth beneath their caps.

In the 19th century, Swedish mycologist Elias Magnus Fries produced a novel system of classification that concentrated upon the colour of spores. Rather than looking at the spores with a microscope, Fries scrutinized 'spore prints' that accumulate when a mushroom cap is placed gills downward on a piece of paper.

After Darwin elucidated the mechanism of evolution, mycologists, and other biologists, attempted to create natural systems of classification in which groups of species are related by descent from a common ancestor. It turns out that mushroom shape, size, and colour are poor guides for natural classification.

As we saw in Chapter 1, genetic comparisons are used to tease apart the largest lineages of organisms into supergroups (e.g. Opisthokonta) and kingdoms (e.g. Fungi and Animalia). The sequences of genes also serve as the most objective guide for identifying individual species and unravelling evolutionary relationships between species. The term 'taxonomy' describes the identification and naming of organisms and 'systematics' refers to the particular study of evolutionary relationships among organisms. The fragment of the complete set of genetic instructions of a fungus that is sequenced for these evolutionary studies is called the ITS region, where ITS stands for Internal Transcribed Spacer. ITS is a region of the gene that encodes part of a cell structure called the ribosome. Ribosomes are the molecular machines that carry out protein synthesis. They are about one million times smaller than cells: a single yeast cell contains 200,000 ribosomes. Mycologists who specialize in the study of fungal systematics amplify all or part of the ITS region from fungi, determine their sequences, and compare the sequences among different fungi to determine their similarity. The ITS region has proven so useful for identifying fungi that the sequence is used in diagnostic kits for the rapid detection of fungi causing human infections, plant disease, and contamination of water-damaged buildings.

Data produced by genetic comparisons is used to construct evolutionary or phylogenetic trees that display relationships between organisms in the form of branching diagrams. Phylogenetic trees are used as a guide for making sensible decisions about grouping species into a Genus, genera into a Family, families into an Order, and orders into a Class (Table 1).

Table 1. Taxonomy of three fungal species

Taxonomic rank			
Kingdom	Fungi	Fungi	Fungi
Phylum	Chytridiomycota	Zygomycota*	Basidiomycota
Class	Chytridiomycetes	Zygomycetes*	Agaricomycetes
Order	Chytridiales	Kickxellales	Geastrales
Family	Olpidiaceae	Kickxellaceae	Geastraceae
Genus	*Olpidium*	*Spirodactylon*	*Sphaerobolus*
Species	*brassicae*	*aureum*	*stellatus*

*Informal ranks

With each step up in this hierarchy of names, we are dealing with greater genetic variation. Related classes are arranged into a Phylum and this is a convenient level for further consideration of the modern classification of the Kingdom Fungi (Table 2). There are some disagreements about the number of phyla and this organization of species is likely to change as more fungi are discovered and more detailed genetic research is completed.

More than 90 per cent of the more than 70,000 species of fungi that have been described by mycologists are classified within Phylum Basidiomycota (basidiomycetes) and Phylum Ascomycota (ascomycetes). Half of the basidiomycetes produce mushrooms; the others include rusts and smuts that cause plant disease, and a plethora of single-celled yeasts. Ascomycetes include the yeast *Saccharomyces cerevisiae*, fungi with beautiful cup-shaped fruit bodies, truffles, and morels. Basidiomycetes produce basidiospores and ascomycetes produce ascospores (Figure 10). The number of technical terms used to describe fungi can be an impediment to learning about them. An imposing book titled

Table 2. Major groups or phyla of fungi

Phylum	Common name	Examples
Basidiomycota	mushrooms	gilled mushrooms, boletes, bracket fungi, jelly fungi, artillery fungus
Ascomycota	cup fungi	morels, truffles, baker's yeast, moulds (*Aspergillus, Penicillium*)
Glomeromycota	arbuscular mycorrhizal fungi	*Glomus* species
Zygomycota (informal)	bread moulds	*Mucor, Rhizopus, Spirodactylon*
Chytridiomycota	chytrids	*Batrachochytrium* (frog pathogen)
Blastocladiomycota	no common name	*Allomyces*
Neocallimastigomycota	anaerobic rumen fungi	*Neocallimastix*
Cryptomycota	hidden fungi	*Rozella*

The Dictionary of the Fungi has been published since the 1940s and lists 21,000 names of fungi and terms that describe their structure. Reliance on this terminology is as limited as possible in this short book without compromising the accuracy of the science.

Mushrooms and related fungi

The apparent stillness of a mushroom belies furious microscopic activity beneath its cap. Gill development provides a mushroom with a twenty-fold larger surface area for spore production than a flat disc of the same diameter. Cells that produce spores are packed on to the gill surface and each spore is launched into the

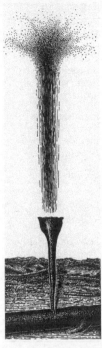

10. **Fruit bodies of (left) an ascomycete cup fungus and (right) a mushroom-forming basidiomycete. The cup fungus discharges ascospores from multiple asci in synchrony producing a vertical puff. The mushroom discharges basidiospores from its gills, which fall from the cap and are dispersed by wind.**

space between neighbouring gills by the motion of a tiny droplet of fluid. This intricate discharge mechanism, called a surface tension catapult, can be observed using a high-speed video camera attached to a microscope. Frame-by-frame viewing of videos captured at 100,000 frames per second shows that the fluid drop slaps on to the spore, sending spore and drop flying from the gills (Figure 11). This happens tens of thousands of times every second, allowing a big mushroom to release three billion spores in a day. Large bracket fungi, which are also basidiomycetes, can release

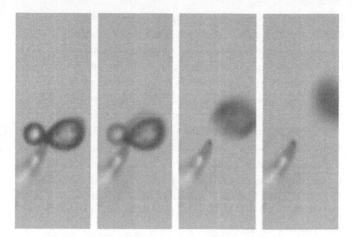

11. Basidiospore discharge shown in successive images from high-speed video recording captured at 100,000 frames per second. The fluid droplet at the base of the spore in the first frame coalesces with fluid on the adjacent spore surface in the second frame, which makes the spore jump into the air at an acceleration of 10,000 g.

trillions of spores every year amounting to a weight of one kilogram.

Umbrella-shaped mushrooms with gills are a common type of basidiomycete fruit body that has developed many times during the evolutionary history of this group of fungi. The repeated emergence of the same basic structure is an example of convergent evolution. This is probably explained by the efficiency of this architecture. Growth of a mushroom draws upon the resources of a large volume of filamentous hyphae that are feeding in the surrounding soil or rotting wood. The development of a mushroom from the colony is an investment in the future of the genes carried by the clouds of spores that will be shed from its gills.

Mushroom stems and caps are fashioned by millions of branching hyphae. The elongation of the stem allows a fruit body to poke a few centimetres above the ground, or to sprout from a tree, placing the gills in a perfect position for misting the air with spores. Cap formation is also important because it protects the gills from raindrops that would sabotage the mechanism of spore discharge. Given the energy consumed in making a mushroom, it makes sense to release as many spores as possible. Alternatives to gilled mushrooms include fruit bodies with pendulous spines (*Hydnum repandum*, the hedgehog mushroom, is an example), and others with tightly packed tubes (boletes like the edible cep or porcini produce these 'poroid' mushrooms). A variety of basidiomycetes dispense with the protection offered by a cap and release their spores from exposed surfaces when it is not raining. These include coral fungi, with fruit bodies that look like tiny candelabra, jelly fungi, which look like jelly, and various species that form flat crusts on tree bark.

Wind dispersal of spores from mushrooms is limited to short distances in sheltered locations and insects are thought to assist in spreading spores of some species. Spores stick to the bodies of insects that feed on mushrooms and are carried in their guts. A bioluminescent mushroom species that grows in the Brazilian rainforest lures insects by emitting green light during darkness. This was demonstrated by experiments using acrylic models of fruit bodies illuminated with green light-emitting diodes. It is not known whether other bioluminescent mushrooms work in this fashion, but insect dispersal is probably important for luminescent and non-luminescent mushrooms.

An assortment of basidiomycetes have abandoned 'active' spore discharge entirely in favour of splash discharge by raindrops and dispersal by insects and other animals. Earth-stars form lovely fruit bodies that enclose spores in round bags that are held above the ground on star-shaped platforms. The bags are compressed by raindrops, expelling a jet of spores through a nozzle at the top. After release, the spores drift away on the breeze. Puffballs and earth-balls work in a similar fashion, although some of them split apart and release spores without using a nozzle to concentrate and accelerate their emissions. The giant puffball, *Calvatia gigantea*, is the most fecund of fruit bodies, with the largest specimens holding an estimated seven trillion spores. Arranged in a line, these 'dust' particles would ring the equator.

Bird's nest fungi produce very elaborate fruit bodies (Figure 12). These basidiomycetes house their spores inside capsules that nestle in their small cup-shaped fruit bodies. Looking from above, the capsules look like eggs in a nest. Each bird's nest capsule is bigger (2 mm or more in diameter) than the single capsule discharged by the artillery fungus (less than 1 mm in diameter). The fruit bodies direct falling raindrops into the bottom of the

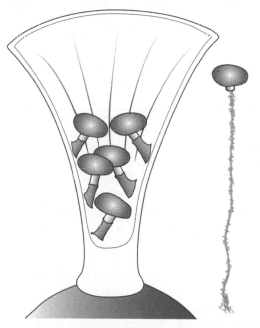

12. **Structure of the fruit body of the bird's nest fungus *Cyathus striatus*.** The interior of the fruit body is shown in this illustration to reveal the capsules before they are splashed from the cup. A single capsule is shown on the right with a fully extended grabline.

nest, splashing their capsules into the air over a distance of up to one metre. Some species have a cable attached to each capsule that operates as a grabline, attaching the capsules to plant stalks as they fly from the fruit body. This tethering mechanism increases the chance that the capsules will be consumed by herbivores that graze around the bird's nest fungi. After passage through the gut of the animal, the 100 million spores inside each capsule will be deposited in dung that serves as a fertilizer for the development of a new mycelium.

Stinkhorns are stranger still. The common stinkhorn, *Phallus impudicus*, is shaped like an erect human penis and attracts flies to the stinking olive-green slime that covers its head. It emerges from a buried 'egg' that looks like a golf ball. The basidiospores of this fantastic organism are embedded in the slime, so that insects consuming the slime transport the spores when they fly away. This reproductive behaviour is comparable to the use of the odour of rotting flesh by corpse flowers to attract the scavenging flies and beetles that act as their pollinators. Relatives of the common stinkhorn form thin red stalks tipped with slime (dog stinkhorn), hang a white network beneath the head that looks like a skirt (veiled stinkhorn), and spread their slime over the 'bars' of cages (latticed stinkhorn) and the 'arms' of red stars (starfish fungus). Each of these odd structures emerges from its egg stage in a few hours by absorbing water. This rapid inflation mechanism is seen to a greater or lesser extent in other mushrooms and accounts for the sudden appearance of fungi that people find so surprising.

Molecular phylogenetic research shows that earth-stars, puffballs, earth-balls, bird's nest fungi, and stinkhorns evolved within different groups of mushrooms with gills and tubes beneath their caps. The loss of the active mechanism of spore discharge was a critical step in this process. The effectiveness of these alternative strategies for spore dispersal is evident from the discovery of fossil representatives of earth-stars and bird's nest fungi which are tens of millions of years old.

The rusts and smuts are classified as separate subgroups of the basidiomycetes. Similarities in cell structure between rusts and smuts and the mushroom-forming species unite them within the basidiomycetes. Their evolutionary ties are also apparent from their use of the droplet mechanism of spore discharge. Rusts are specialized plant parasites that attack a wide range of plants including wheat, coffee, and other critical agricultural species. Smuts infect the reproductive organs of flowering plants. In loose smut of barley, for example, a smut fungus infects the open flowers and the damaged flower heads fill with masses of black spores rather than seeds.

Ascomycetes and other groups

With a few exceptions, ascomycetes are overlooked more easily than basidiomycete mushrooms. With the exception of morels, truffles, and a handful of rarer species, the beauty of ascomycete fruit bodies is invisible without the aid of a hand lens. The inconspicuous nature of these fungi is no excuse for ignoring them. Ascomycetes are the most important pathogens of plants, cause food spoilage, are a major cause of human allergies, and include the yeast that is our dependable workmate in biotechnology. And, outside the most polluted cities, we see ascomycete colonies everywhere in the form of the fungal components of lichens.

The ascospores of ascomycetes are formed inside cells called asci. Lacking the drop mechanism of spore discharge used by basidiomycetes, many ascomycetes shoot their spores from asci that operate as pressurized cannons. Ascospore discharge is one of the fastest movements in nature, with a record speed of 100 kilometres per hour, measured using a high-speed video camera running at one million frames per second (Figure 13).

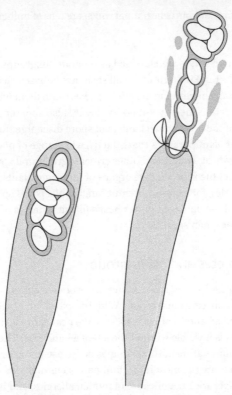

13. Ascospore discharge in *Ascobolus immersus* whose asci open via a lid. The spores of this fungus are connected by mucilage and the whole mass is shot at a speed of 65 kilometres per hour. The fastest of the ascomycetes propel their spores at speeds of up to 100 kilometres per hour.

Some ascomycetes produce 'naked' asci. *Taphrina deformans*, that causes leaf curl disease in fruit trees, spreads itself from asci that break through the surface of infected leaves and spray their spores into the air. Ascomycetes in the genus *Dipodascus* form long asci on surfaces and exude their spores through the expansion of mucilage within the ascus. And hundreds of species of ascomycete yeasts transform their single cells into asci and spill their ascospores when the walls of their asci dissolve. But the common name of the phylum, the 'cup fungi', refers to a type of ascomycete fruit body called an apothecium.

Apothecia range in size from tiny blue discs of the bluestain fungus, *Chlorociboria*, and red discs of the eyelash cup fungus, *Scutellinia scutellata*, to the pale brown pig's ear fungus, *Peziza badia*, that can grow as big as a cereal bowl. The fruit bodies of edible morels and toxic false morels are examples of big apothecia that are pushed into the air on stems like basidiomycete mushrooms. Truffles are modified apothecia whose exposed surfaces have become enclosed during the evolutionary history of these fungi. They form below ground where they use pheromones to attract mammals that consume the fruit bodies and disperse the ascospores in their dung. Other kinds of ascomycete fruit body develop as tiny flasks (perithecia) and balls (cleistothecia). *Neurospora crassa* is a perithecial ascomycete that is a favourite for cell biological research, and species of *Aspergillus* produce cleistothecia.

The fruit bodies of ascomycetes, like those of basidiomycetes, are sexual organs that release spores that result from sexual interactions between mycelia of compatible mating type. Ascospores and basidiospores are sexual spores. The life cycles of many of these fungi—particularly the ascomycetes—are complicated by the production of other types of spores without any sexual process. These asexual spores, or conidia, are produced by single colonies and this mechanism of asexual reproduction is a very successful way for spreading the fungus (Figure 14). Depending upon food availability and other environmental conditions, and the proximity of compatible mates, a single ascomycete species can cycle between asexual reproduction using conidia and sexual reproduction using ascospores. This is called pleomorphism or pleomorphy. It creates problems for describing fungal species because some fungi have been named twice, once for the conidium phase, and a second time for the ascospore phase.

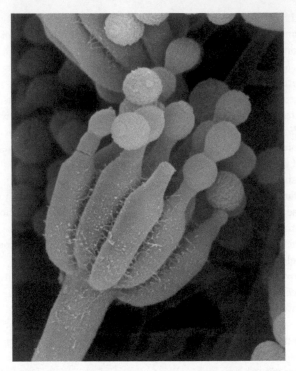

14. Asexual spores (conidia) of a *Penicillium* species produced from a cluster of elongated cells at the top of a stalk (conidiophore).

This has resulted in the proliferation of Latin names for fungi and taxonomists try to cancel duplicate names when they are identified. We will consider other problems in taxonomy shortly.

The other major groups of fungi are less well known. They contain fewer described species than the basidiomycetes and ascomycetes, and few of them are visible without a microscope. Phylum Glomeromycota is a group of fungi that forms supportive relationships with plants called arbuscular mycorrhizas. The term 'arbuscular', meaning 'dwarf tree', refers to the finely branched structures that the fungus forms inside the root cells of its plant partner. Arbuscules provide a large surface area for the exchange of nutrients and water: the fungus supplies the plant with water and dissolved minerals and the plant provides the fungus with some of the sugar that it produces by photosynthesis. Zygomycete fungi used to be classified in a separate phylum, but recent research shows that these microorganisms are scattered across the evolutionary tree. They do not form a coherent natural group. Mating between two strains of the zygomycete *Mucor* is often used as a textbook example of a simple form sexual reproduction. *Pilobolus* (Figure 6) and *Spirodactylon* (Figure 8) are also zygomycetes. Other zygomycete fungi are important in food spoilage, are used to ferment soybean to produce tempe, and can cause serious human infections called mucormycoses.

Four phyla of fungi produce swimming spores called zoospores. The names for these phyla are Blastocladiomycota, Chytridiomycota, Neocallimastigomycota, and Cryptomycota. *Batrachochytrium dendrobatidis* is a chytrid associated with the global decline in amphibian populations. Its zoospores infect frogs and other amphibians, producing cysts in their skin that develop into spherical thalli from which the next generation of swimming spores are released. Other chytrids, as well as some of the Blastocladiomycota, produce elongated thalli that release zoospores from sporangia that develop at the tips of branching filaments. Species of Neocallimastigomycota grow in the digestive systems of cows and other herbivores. They are strict anaerobes, poisoned by oxygen, and aid the digestion of plant materials in their hosts. The Cryptomycota, introduced in Chapter 1, have been identified in all kinds of aquatic habitats using metagenomic methods. Some of them infect other fungi and diatoms, but we know very little about these microbes because we have not been able to grow them in culture.

The evolutionary tree of the fungi becomes very complicated when we consider the zoosporic species and it is difficult to be clear about which species fit into which group and whether some should be regarded as fungi at all. Although *Olpidium* (Figure 7) looks a lot like *Batrachochytrium*, genetic comparisons show that it is a closer relative of the zygomycetes. So, despite its simple structure and zoospore formation, the classification of *Olpidium* as a chytrid is questionable. The position of microorganisms called Microsporidia is another problem for taxonomists. Microsporidia are parasites that grow inside the cells of insects, crustaceans, and fish. Some of them cause intestinal infections in people with damaged immune systems. Like many parasites, their genomes have shrunk as they have become more and more reliant upon the activities of their hosts. Microsporidia are related to 'basal' fungi, meaning that they seem to be affiliated with the ancestors of the fungi, but it is arguable whether they should be classified as fungi or as a group outside the fungi. This lack of clarity is not surprising because all groups of organisms bleed into one another through their ancestral roots.

Whether or not microsporidia are proper fungi, the study of mycology encompasses a tremendous range of organisms. More than 16,000 kinds of mushroom-forming basidiomycetes have been described and the real number is probably twice as big. For comparison, the catalogue of vertebrate animals is close to complete and zoologists recognize 5,400 mammals and 10,000 birds. Mycologists who are dedicated to the identification of fungi have a lot more to contend with than birdwatchers. A related technical challenge in fungal biology is the problem in defining species. Separate species of birds are characterized by the reproductive compatibility between males and females within the species, their reproductive isolation from other birds, and a suite of visible (morphological) and genetic characteristics. Even in ornithology, however, the spread of genetic variation across several subspecies of a bird and the birth of hybrids between species introduce uncertainty in species descriptions. The challenges are bigger in mycology because many fungi will not reproduce sexually in culture and so many of them look the same. This is the reason that new species descriptions rely so heavily on genetic studies.

In the 20th century it was not unusual for a professional mycologist with expertise in taxonomy to spend an entire career describing the microscopic features of a small group of fungi. Shelves of academic monographs have been published about the species within a single fungal genus. This may seem futile until we consider that a monograph on the ascomycete genus *Fusarium* serves as a guide to identifying a growing list of species that decompose plant debris in soils, cause destructive plant diseases, produce toxic compounds called mycotoxins, and cause human infections. With growing molecular exploration, a lot of this know-how is being lost as the survey of life reveals ever greater variety at the level of genetics. There is, however, no substitute for looking at fungi in nature to appreciate their spectacular diversity. With the company of an experienced guide, a morning walk in the woods after a period of heavy rainfall can provide an inspiring introduction to the fungi. Thanks to macro photographers who celebrate the beauty of the fungi in online collections of colourful images, the armchair mycologist can also be entertained and informed without getting muddy.

Chapter 3
Fungal genetics and life cycles

Genomes

All of the structures produced by fungi, from simple budding yeast cells to long-lived bracket mushrooms, are encoded in genes. The genome of a fungus is a blueprint for its organization and operation. Describing genes as components of a 'blueprint' may seem too simple when we think about something as complicated as a living organism. But there is no other type of stored information that is transmitted from one generation to the next, ensuring that black truffles, for example, will develop from the spores of black truffles. This does not mean that every fungus with black truffle genes will look exactly the same. Black truffles come in many shapes and sizes, ranging from wrinkled nuggets the size of walnuts to giants that weigh more than one kilogram and are a sensation at truffle auctions. Truffles with certain versions of genes may be better at getting big, but most of the differences in size and shape are caused by variations in soil type, temperature, rainfall, and the vitality of the truffle colony. Because truffles are mycorrhizal fungi, their development is also influenced by the health of the hazel and oak trees to which they are connected.

The genome of the Périgord black truffle, whose Latin name is *Tuber melanosporum*, is huge for a fungus. There are a number of ways to compare genomes of different organisms. If we look at the number of pairs of the letters A, T, G, C (nucleotides) organized in the DNA in the cell, we find that the truffle genome is ten times larger than the genome of the yeast, *Saccharomyces cerevisiae*, and four times the size of the genome of the cultivated button mushroom, *Agaricus bisporus*. This measure of genome size can be a poor indication of the number of functional genes that specify proteins: yeast has 6,000 genes, the button mushroom has 10,000 genes, and there are 7,500 genes in the truffle genome. The enormity of the truffle genome is explained by the incorporation of a great deal of non-coding DNA—DNA that does not code for proteins—whose function is unclear. Much of this is in the form of repetitive DNA sequences, including transposable elements that copy themselves and move around the genome.

The sequencing of a genome is one of the most important steps in ongoing research by biologists to understand the mechanisms that control the development and functions of organisms. The second phase of this exploration is to distinguish genes that encode proteins and determine the nature of the proteins encoded by individual genes. This is the process of genome annotation, which is considered part of the research field called bioinformatics. Investigators have identified genes in the black truffle genome that control the production of the flavours and volatile aromatic compounds that rodents, truffle flies, and human truffle lovers find so seductive. Annotation of the truffle genome has also revealed genes that control carbohydrate metabolism and sexual reproduction. Elsewhere in the genome, waiting to be found, are genes that manage the chemical communication between the fungus and the roots of its plant partner, genes that trigger fruit body formation, and genes that shape the beautifully ornamented spores of the truffle.

Yeast

Genetic research is much more advanced in the study of the ascomycete yeast *Saccharomyces cerevisiae*. There are many reasons why such spectacular progress has been made in understanding how this microorganism works. The most important of these is the ease with which yeast can be grown in culture. Cultures are started by spreading a tiny quantity of yeast cells on agar using a wire inoculating loop, or by transferring cells to liquid growth medium (broth) using a plastic pipette. When the cultures are incubated at 30°C the yeast cells divide by forming buds every 1–2 hours, so that a single yeast produces a group of sixteen cells in less than eight hours, and more than 4,000 cells in a day. Truffles, and many other fungi, do not lend themselves to this kind of straightforward manipulation in the lab. Truffle growers have to wait for seven to fifteen years after inoculating trees before the first harvest. Anyone committed to experiments on truffles must be extraordinarily patient.

Genetic research on *Saccharomyces* began in the first half of the 20th century. As techniques improved, yeast was adopted as a favourite experimental organism for studies on molecular genetics, cell biology, and evolution. Yeast became a 'model' eukaryote in the same way that the gut bacterium *Escherichia coli* serves as a model prokaryote. To understand how the genetics of yeast are manipulated in the laboratory we need to look at the process of yeast reproduction. Budding yeast cells come in two strains called **a** and α. These mating types are akin to sexes, although there is no way to tell them apart from their appearance under the microscope. The difference between **a** and α cells lies in the production of different chemical attractants, or pheromones, and complementary receptor molecules in their membranes. When cells of the opposite mating type detect each other's pheromones, their surfaces bulge, make contact, and fuse to produce a larger combined cell that contains two complete sets of chromosomes. This is similar to the fertilization of an egg by a sperm in animal biology. In both cases, two cells with single sets of chromosomes combine to form a single cell with two sets of chromosomes. Cells with a single set of chromosomes are called haploid; cells with two sets are called diploid. Life cycle diagrams are helpful in illustrating how the switch between the haploid and diploid state relates to the development of different organisms (Figure 15).

The diploid yeast cells can bud, but when food becomes limited they engage in a special kind of cell division called meiosis that produces four haploid cells each containing a single set of chromosomes. These haploid cells are called ascospores. They sit inside the diploid cell, which is called the ascus, where they develop a thick cell wall. Yeast ascospores function as survival capsules that allow the fungus to wait for a change in

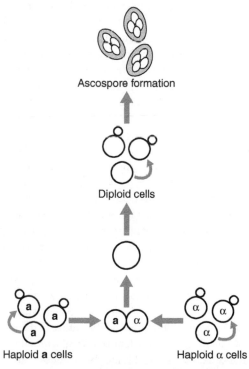

Ascospore formation

Diploid cells

Haploid **a** cells   Haploid α cells

15. **Sexual reproduction in baker's yeast, *Saccharomyces cerevisiae*.**

environmental conditions. The group of four ascospores is called a tetrad. Tetrads are interesting from a genetic point of view because each of the new spores contains a unique version of the yeast genome. These exclusive genomes originate from the original fusion of two cells ($\mathbf{a} \times \alpha$), followed by the rearrangement of genes during meiosis. The same process of genetic recombination in humans produces the differences between sperm and egg cells that result in the profound differences between children, their siblings, and their parents.

To determine the function of a particular gene, researchers work with mutant strains of yeast that carry mutant versions of the gene. Natural mutations are caused by ionizing radiation, oxidizing agents produced in cells, and errors during DNA replication. Mutant strains are produced in the laboratory by treating cells with chemical mutagens or by exposing cells to ultraviolet light. Both treatments damage DNA structure, producing mutations in a random fashion throughout the genome. Site-directed mutagenesis offers greater precision by introducing mutant copies of specific genes into yeast chromosomes.

Tetrad analysis is a technique used to determine whether or not a particular strain of yeast is carrying a single mutation. Tetrad analysis is also used to map the positions of genes on the sixteen chromosomes in yeast. Tetrads of ascospores are produced by crossing cells of the two mating types and researchers separate the four ascospores from each tetrad into spots arranged in a grid on a culture plate. Between ten and twenty tetrads are processed in this way on a single culture plate, producing forty to eighty spots occupied by single spores. Ingredients added to the growth medium in the plates are used to select for cells with a particular genetic makeup. The growth patterns that develop when the plates are incubated provide information on the nature of the mutant gene. If, for example, two of the four cells from a tetrad form colonies and two do not grow, this may be evidence that yeast is disabled by the mutant version of the gene. This encourages

Fungi

further experiments to determine how the normal gene operates. The conventional technique for separating spores from a tetrad uses a microscope and glass dissection needles controlled with a device called a micromanipulator. Skilled investigators can separate up to sixty tetrads in an hour, but this is exceedingly tedious work. In recent years, the method has been simplified with the automation of some of the steps. Researchers are also working on a hands-free technique that sorts cells using a flow cytometer that reads a unique genetic barcode introduced into each tetrad.

Another important method in yeast genetics is called two-hybrid screening. This utilizes a reporter gene within an engineered strain of yeast to identify interactions between different proteins encoded in the genome. The study of protein–protein interactions is crucial for understanding the function of different genes and their protein products. The combination of decades of research on yeast and development of increasingly powerful methods in molecular genetics and cell biology has enabled investigators to learn more about the way that yeast works than any other eukaryote. This does not mean, however, that we understand everything about this single-celled fungus. Even though its genome was published in 2001, the function of more than 10 per cent of the genes in *Saccharomyces* is unknown. Fission yeast, *Schizosaccharomyces pombe*, is another single-celled ascomycete that has been studied in great detail. Its genome was sequenced in 2002 and encodes almost 5,000 genes. When fission yeast divides, an internal septum grows across the midpoint of the cell producing two identical 'daughter cells'. This is completely different from budding in *Saccharomyces*. The molecular control of cell division in fission yeast has significant implications for cancer research.

Filamentous fungi

The genetics of filamentous ascomycetes is not understood in as much detail as yeast genetics, but a few species have become important experimental organisms. *Neurospora crassa* has been studied by geneticists since the 1940s and was used in experiments that linked gene expression to the synthesis of enzymes. The life cycle of *Neurospora* is more complicated than yeast. Like yeast, there are two mating types. Instead of the direct fusion of growing cells that occurs in yeast, *Neurospora* uses airborne spores called microconidia to fertilize cells on the surface of developing fruit bodies. The fruit bodies of *Neurospora* are called perithecia. Fertilization creates cells that contain nuclei of both mating types. Fusion of a pair of these nuclei followed by meiosis produces ascospores. In *Neurospora*, meiosis is followed by a mitotic division so that groups of eight spores are formed rather than the tetrads of *Saccharomyces*. These spores are arranged in single file inside tubular asci that extend through a hole in the top of the perithecium and blast them into the air above the fungal colony. The perithecia are less than 0.5 millimetres in width and contain up to 300 explosive asci. (The function of asci as pressurized cannons was introduced in Chapter 2.) The *Neurospora* genome contains 10,000 genes that encode proteins. The large number of genes relative to yeast reflects the formation of multicellular fruit bodies and the development of the branched colonies of filamentous hyphae during the feeding phase of the life cycle.

Basidiomycete fungi that produce mushrooms engage in cell fusion and meiosis to form spores like the ascomycetes, but there are some fundamental differences in the mating processes in these groups. In mushrooms, the fusion of cells occurs long before nuclear fusion (fertilization), meiosis, and spore formation. Colonies keep growing after merger and delay the processes of recombination until the combined organism has formed a fruit body. The details vary from species to species. Feeding mycelia of the shaggy mane or lawyer's wig mushroom, *Coprinus comatus*, are common inhabitants of garden lawns. Hyphae that form these mycelia are divided by septa into compartments that contain a single nucleus. Hyphae of this kind are called homokaryons.

When hyphae of compatible strains meet in the soil their cell walls fuse to produce a heterokaryon in which each compartment contains one nucleus from each homokaryon (Figure 16). This condition with two nuclei per compartment is maintained by the formation of tiny branches on the outside of the cell, called clamp connections. Each clamp connection acts as a bridge that allows nuclei to flow between compartments. Some books refer to homokaryons as monokaryons and heterokaryons as dikaryons.

Sexual compatibility in the shaggy mane mushroom is controlled by a single gene or mating type locus. This is known as a

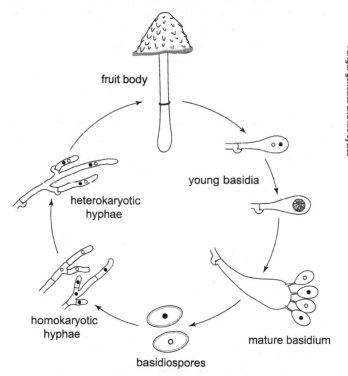

fruit body

young basidia

heterokaryotic hyphae

mature basidium

homokaryotic hyphae

basidiospores

16. The life cycle of the mushroom *Coprinus comatus*.

bipolar or unifactorial incompatibility system. Complexity is introduced to this reproductive mechanism by the presence of different versions, or alleles, of the mating type locus (A_1, A_2, A_3, etc.). When mycelia with the same mating type allele make contact they are incompatible (e.g. $A_1 \times A_1$). But whenever a shaggy mane mycelium meets any mycelium with one of the other mating type alleles it has the potential to form a heterokaryon (e.g. $A_1 \times A_3 = A_1 A_3$). This favours outbreeding and maximizes the probability of sexual reproduction when colonies encounter one another. The more complicated tetrapolar incompatibility system relies on a pair of genes that control mating. A successful cross between two homokaryons, $A_1 B_1 \times A_2 B_2$, produces a heterokaryon carrying both kinds of nucleus: $A_1 A_2 / B_1 B_2$. Many different alleles of both genes have evolved in some mushrooms, so that the number of possible combinations of homokaryotic mycelia skyrockets. In another ink cap mushroom, *Coprinopsis cinerea*, there are tens of thousands of different mating types based on this tetrapolar system (e.g. $A_1 B_1$, $A_1 B_2$, $A_1 B_9$, $A_2 B_7$, and so on). This reproductive system seems mind-boggling when compared with the binary approach employed by our species. It makes sense for mushrooms whose survival relies on chance engagements with a colony of the same species growing in the same patch of soil. *Coprinopsis cinerea* is a model for mushroom research. Its genome encodes around 13,000 genes. The genomes of many other mushrooms are considerably larger.

Once a heterokaryon forms, mushrooms develop according to an internal clock if soil moisture, temperature, and other environmental conditions are permissive. Mushroom development is a mysterious process. Mushroom formation begins with the growth of a knot of hyphae that expands into a rounded primordium. If the primordium is sliced from top to bottom an embryonic fruit body is often evident with miniature cap, gills, and stem. In the poisonous death cap, *Amanita phalloides*, the primordium develops into an egg-shaped structure in which the mushroom is surrounded by a protective layer of hyphae called the universal veil. As the mushroom expands, the universal veil is stretched and broken and the lower part of it persists as a cup called the volva that surrounds the bulbous base of the stem (Figure 17). Fragmentation of the upper part of the universal veil leaves white spots on the red cap of the fly agaric, *Amanita muscaria*, and scales on the caps of many other mushrooms. A second interior sheet of hyphae, called the partial veil, protects the gills in the primordium. Breakage of this membrane during cap expansion leaves a ring around the top of the stem in the death cap and the fly agaric.

The overnight appearance of mushrooms is one of the phenomena that have encouraged so much superstition about fungi. It is explained by the unusual way that fruiting occurs. Rather than relying on an increase in the number of cells in the fruit body, most of the rapid increase in the volume of mushroom is caused by the absorption of water and elongation of the millions of hyphae that form the stem and the cap. This hydraulic mechanism allows the fruit body to exert a pressure as it thrusts up through soil or rotting wood. In the urban environment, this exertion of force explains

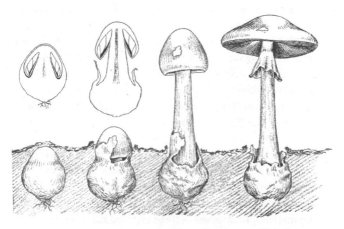

17. **Mushroom expansion in *Amanita phalloides*, the death cap.**

how clusters of mushrooms push their way through crevices in concrete and are even capable of rupturing a layer of asphalt.

While experiments have allowed mycologists to solve these mechanical aspects of mushroom development, we have few clues about the genetic controls of fruit body development. We do not know how the genome of the death cap always produces mushrooms with greenish or yellowish caps, evenly spaced white gills, and solid stems with a ring and a volva. Questions about the sensitivity of mushrooms to gravity are unanswered. The gravitropic growth of mushrooms aligns the cap horizontally and the gills vertically. This is essential to ensure that discharged spores fall between gills and escape into the air rather than becoming stuck on a gill surface. The way that mushrooms sense gravity is a mystery.

There is so much to learn about mushrooms, so many prospects for young investigators. The mushroom life cycle has been studied for more than one hundred years and the genetic control of mating remains an active research topic. As a young doctoral student, Elsie Wakefield (1886–1972) performed crosses between different mushroom strains to explore the genetics of the mating process. She went on to become Head of Mycology at Kew in London and is one of many women scientists who have dedicated their careers to mycology. Mycology is one of the branches of biology that has always attracted women and it is important to note Elsie Wakefield's work to balance the overwhelming influence of Arthur Henry Reginald Buller (1874–1944), who is recognized as 'The Einstein of Mycology'.

Buller was born in Britain and spent his career at the University of Manitoba in Canada. His seven-volume work, *Researches on Fungi*, is the foundation of the modern study of mycology. These books describe ingenious experiments on many aspects of mushroom biology and some of them exposed his eccentric personality. When he worked on bioluminescence, for example, he strapped leather horse 'blinders' to his head to avoid exposing his eyesight to bright street lamps. This strategy allowed him to preserve his sensitivity to the dim light thrown by mushrooms in his laboratory. He also made significant contributions to fungal genetics including a prediction, which was validated after his death, that outcrossing in mushrooms could occur through the transmission of nuclei from a heterokaryon into a homokaryon. This esoteric point is an important feature of mushroom biology and has practical applications in mushroom cultivation.

Buller worked on many kinds of fungi, including rusts that infect cereal crops. Rusts are examples of basidiomycete fungi that do not form mushrooms. They have the largest genomes of any fungi, with one species packing more than seventy times more raw sequence than yeast and encoding over 25,000 genes. For further comparison, there are an estimated 20,000 to 25,000 genes in the human genome. The reproductive mechanisms of the rusts will be discussed in Chapter 5 when we consider fungi that cause plant diseases.

Genetic research on the arbuscular mycorrhizal fungi (Glomeromycota), zygomycetes, and aquatic fungi that produce swimming zoospores is very limited in comparison with the investment made in understanding the reproductive biology of ascomycetes and basidiomycetes. Part of the explanation for this research bias lies in the difficulty of working with these fungi in culture. Fungi in the Glomeromycota that form arbuscular mycorrhizas with plants do not grow on their own in culture at all, but they can be grown in association with cultured roots. Asexual reproduction in these fungi is accomplished by the formation of huge spores. Some have a diameter of almost 1 mm and are filled with hundreds or thousands of nuclei. They are formed in clusters called sporocarps in some species. The genome of *Rhizophagus irregularis*, a fungus that forms arbuscular mycorrhizas with poplar trees, is even larger than the truffle genome and encodes 28,000 genes. Like the truffle genome, the chromosomes of *Rhizophagus* incorporate multiple copies of genes. It is possible that the entire genome of this fungus has been duplicated during its evolution. A lot of the protein-encoding genes seem to be involved in the communication between the fungus and its plant host. Sexual reproduction has not been observed in arbuscular mycorrhizal fungi, but it may occur in nature. Evidence for this is found through annotation of the *Rhizophagus* genome, which reveals that the fungus has genes that control meiosis and genes that specify different mating types.

Zygomycetes produce sexual spores called zygospores. Zygospores resist germination in culture, which makes it very difficult to assess the outcome of crosses between different strains. Sexual reproduction in these fungi has been studied in detail, however, and is often used as an illustration of a simple life cycle in biology textbooks (Figure 18). Mycelia of these fungi, including species of *Mucor* and *Rhizopus* (not to be confused with *Rhizophagus*), are not separated into compartments by septa. Their nuclei are distributed throughout their branching hyphae. The sequenced genomes of species of *Mucor* and *Rhizopus* are comparable in size to the genomes of mushrooms.

Zygomycete colonies of compatible mating types use volatile pheromone molecules to detect one another before they make physical contact. When hyphae sense these chemicals they make contact and stick together tip-to-tip. After attachment, each cell

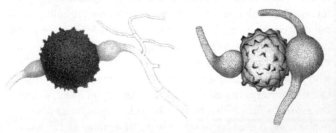

18. **Mature zygospores formed between compatible strains of zygomycete fungi.**

produces a septum behind its tip isolating a compartment that contains multiple nuclei. The two cells merge when the barrier in the contact region dissolves, and a thick wall forms around the blended cytoplasm to produce a zygospore. Nuclei from the two mating types fuse inside this spore and then undergo meiosis to produce a new population of haploid nuclei. These are released in airborne spores when the zygospore germinates and the spores produce a new generation of feeding mycelia. *Spirodactylon*, whose coiled spore-producing stalks were described in Chapter 2, is another zygomycete. Zygospores of this fungus have smooth walls rather than the warty surfaces shown in Figure 18.

Aquatic fungi

Asexual reproduction in aquatic fungi is accomplished by the formation of zoospores. Like the conidia of ascomycetes, zoospores carry clonal copies of the genome of single parents. Unlike conidia, which are blown around by air currents and dispersed by water, zoospores are capable of swimming towards new sources of food using chemical cues in the environment. The pathogenic chytrid *Batrachochytrium dendrobatidis* uses zoospores to locate and infect its amphibian hosts. Its genome includes almost 9,000 protein-encoding genes, which is considerably more than yeast. Sexual reproduction has not been described in this fungus and there is no trace of genes that might control mating type in its genome.

Sexual cycles do occur in other groups of zoosporic fungi. *Allomyces macrogynus* is a species classified in the Blastocladiomycota that produces two types of swimming sex cells or gametes that resemble its asexual zoospores. One of the gametes is much larger than the other and releases a pheromone called sirenin, named after the sirens that tried to lure Odysseus and his shipmates to their doom in Homer's *Odyssey*. Sirenin attracts the smaller type of gamete and fusion occurs when they make contact. After the development of a diploid stage in *Allomyces*, meiosis restores the haploid number of chromosomes in the following generation. The same shift from haploid to diploid, and diploid to haploid, occurs in all of the sexual life cycles described in this chapter.

It is interesting to compare fungal and animal life cycles. In every version of sexual reproduction in fungi and animals, the merger of two cells, or gametes, and fusion of their nuclei doubles the number of chromosomes, and this fertilization event is followed by meiosis that halves the number of chromosomes. The difference between the reproductive cycles of fungi and animals lies in the way that the gametes are produced. Adult fungal cells have a single set of chromosomes; they are haploid and can function directly as gametes. Adult animal cells have two sets of chromosomes; they are diploid and gametes are produced by meiosis. The reason that sexual reproduction occurs in either group of organisms is that it allows for the reshuffling of the genome during fertilization and meiosis. This spreads different versions of genes (alleles) and different combinations of genes across populations, which is the basis for the origin of new species.

Chapter 4
Fungal mutualisms

Symbiosis

Symbiosis is a catch-all term that describes close biological interactions between two or more species. It embraces damaging relationships as well as life-sustaining coalitions. A bracket mushroom protruding from the trunk of a dead tree may be the last stage in a parasitic symbiosis. The fungus was supported by this symbiosis at the expense of the infected tree. Relationships from which both the fungus and its partner benefit are called mutualisms. Commensalism lies between these extremes, referring to situations in which one player is advantaged without having any discernible negative or positive effects on the other (Figure 19). Trichomycetes are fungi that live inside the guts of invertebrates and benefit from the stable environment and nutrients available in this habitat. They are commensals that do not seem to impact the health of their hosts. This chapter is devoted to mutualisms.

Mutualisms with insects

Scale insects engage in curious relationships with fungi that trap them beneath mats of tissue made from interlaced hyphae

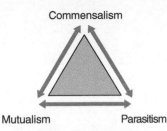

Commensalism

Mutualism   Parasitism

19. Continuum of relationships between different species. Symbiosis is the general term for all of these interactions.

(Figure 20). Female scale insects are limbless plant dwellers that clamp themselves to plants. They feed through slender sucking tubes that penetrate the plant and probe its vascular tissues containing sugary sap. The arrangement between the insect and the plant is like the connection between an oil rig, its drill, and a deep well. The scale insect is a parasite that enjoys all of the nutritional benefit at the plant's expense. Fungi that specialize in relationships with scale insects are basidiomycetes classified in a family called the Septobasidiaceae. There is no common name for this group. The fungus grows around the body of the insect, penetrates its exoskeleton, and feeds on the animal's soft tissues. The fungus seems to be operating as a parasite that lives on a parasite. But rather than harming the insect, the fungus hides its captive from predators in the early stages of the association, allowing it to feed in safety. For this reason, the interaction between the fungus and scale insect is regarded as a mutualism. Rust fungi that cause epidemic plant diseases are close relatives of the Septobasidiaceae. There is no equivocation in identifying rusts as parasites, because their growth does not help the host plant at all.

Ambrosia beetles cultivate fungi in galleries excavated in the wood of trees and shrubs. Mycelia of these ambrosia fungi grow on the walls of the galleries, digesting cellulose and other components of the wood. The beetles enjoy an exclusively fungal diet by consuming cells from the surface of the mycelium. Growth of the ambrosia fungi in the galleries is not left to chance. The beetles

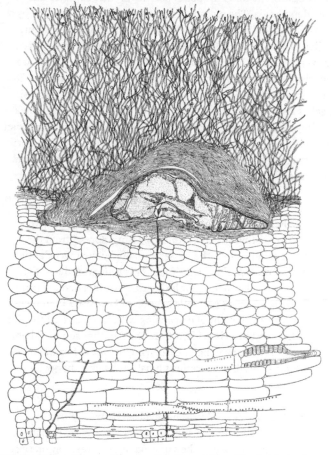

20. Scale insect trapped on a leaf surface within a blanket formed by the fungus *Septobasidium fumigatum*. The insect feeds from the plant through its long sucking tube and the fungus feeds from the insect.

carry cells of the fungi in special pockets in their exoskeletons called mycangia. The mycangia of some beetles are protected by a rim of hairs that brushes fungal spores from the walls of rotting galleries into the pockets. Once inside the mycangia the fungi grow as yeasts rather than filamentous hyphae. These yeasts are nourished by secretions into the mycangia until they are deposited in fresh galleries and begin the process of wood decay. Ambrosia beetles favour damaged and dying trees, but some species attack healthy trees and cause considerable economic damage. The Asian ambrosia beetle is an invasive insect that was introduced to North America in the 1970s. It attacks 200 species of trees, ornamental shrubs, and vines.

Fungus-stealing or 'mycocleptic' ambrosia beetles exploit the work of the 'normal' ambrosia beetles, by tunnelling into wood adjacent to existing galleries. Ambrosia fungi spread from the first set of galleries into the passages made by the mycocleptic beetles and provide them with food. This process reduces the quantity of wood available to the beetles that invaded the trees in the first place. The mycocleptic beetles still excavate their own galleries, but the benefit of dispensing with the active cultivation part of the symbiosis is evident from the fact that some of them no longer produce mycangia. Dispensing with the formation of these organs and their secretions must save some energy for the thieves.

A similarly complex symbiosis has evolved between *Sirex* woodwasps and the fungus *Amylostereum areolatum*. Female woodwasps drill holes into the sapwood of pine trees, lay one egg at the bottom of each hole, and add spores of the fungus on top. Like the ambrosia beetles, *Sirex* woodwasps have special pockets for carrying fungal spores. These mycangia are close to the insect's ovipositor, stationing them for injection into the wood along with the eggs. The wasps also release a dose of toxin carried in mucous that weakens the defence response by the tree. The spores are embedded in wax when they are expressed from the mycangia.

This is dissolved by the wasp's mucous, allowing the fungus to commence growth on the sapwood. When the eggs hatch, the larvae feed on the fungal mycelium. *Amylostereum* is a basidiomycete related to species of *Russula* that produce umbrella-shaped mushrooms. It forms a crusty fruit body on the surface of decaying wood and benefits from its mutualism by its injection through the tough tree bark by the wasp.

Symbioses between fungi and scale insects, ambrosia beetles, and woodwasps are highly developed relationships that have involved substantial structural, biochemical, and behavioural adaptations in the participating organisms. Mutualisms between fungi and social insects reach the height of complexity and are described as a form of mushroom farming. Leaf-cutter ants cultivate mushrooms in underground nests in South America and North America. Millions of insects can occupy the largest nests and support hundreds of fungus gardens that resemble honeycombs. Sterile female workers move between gardens through tunnels and feed the fungi with a pulp made from leaf fragments. Forty ant species engage in this symbiosis across the Americas with a single mushroom species called *Leucoagaricus gongylophorus*. The mycelium of this mushroom digests the leaf pulp using enzymes that convert cellulose and starch into sugars, and break down proteins into amino acids. The ants feed on buds produced on the surface of the mycelium. By depositing faecal fluid containing active enzymes obtained from the buds, the ants jump-start the composting of new leaf material before it becomes colonized by new fungal growth.

The presence of massive quantities of leaf pulp in the ant nests provides an opportunity for the growth of other kinds of fungi. The ants groom their gardens, removing foreign spores and chunks of mycelium infected by these intruders. An ascomycete called *Escovopsis* specializes in attacking the fungal mycelium rather than competing with it for access to the leaf pulp. Spores of this damaging 'mycoparasite' are collected by the ants, pressed into a pocket inside their mouths, and regurgitated outside the nest. Before they are dumped, the spores are sterilized by antibiotics produced by bacteria that grow on the surface of the ants.

The cultivation of fungi is quite widespread among other kinds of ants. One group of ant species farms coral fungi in small nests containing fewer than one hundred workers. Another group cultivates colonies of fungi growing as yeasts rather than filaments. Ants feed coral fungi and yeasts with a mixture of organic materials including wood particles, seeds, plant sap, nectar, dead flowers, and insect excrement. All of the types of ant agriculture evolved from a common ancestor that lived fifty million years ago by farming a variety of fungi, rather than a single mushroom species. Leaf-cutter agriculture is the most recent innovation, which began about ten million years ago.

Termites engage in a comparable form of fungal agriculture in Africa and Asia. Colonies of one or two million insects build mounds of mud and tend mycelia grown on macerated plant matter. Foraging termites chew and swallow grass and wood and deposit this as faeces when they return to the nest. Workers inside the nest fashion this undigested slurry into spongy masses called combs. Mycelia of the symbiotic fungi decompose the plant tissues in the combs, producing sugars and proteins that nourish the colony. The fungi in these relationships are species of *Termitomyces*, including the West African mushroom, *Termitomyces titanicus*, whose enormous gilled fruit bodies have a diameter of up to a metre. Young workers eat fungal nodules on the surface of the combs that could otherwise develop into mushrooms. Once a mound is abandoned, the fruit bodies emerge from the combs and plough their way through the wall. The cathedral mounds built by some termite species can exceed five metres in height. Airflow along a central shaft in the mound and through the porous mud walls maintains a relatively constant temperature that optimizes decomposition by the fungus. The combination of climate control and careful husbandry of the fungus by the termites is comparable to the methods of industrial mushroom cultivation by humans.

Mutualisms with plants: mycorrhizas and endophytes

Mycorrhizas are, seemingly, less lively kinds of fungal symbioses than those associated with social insects. Lacking the visible bustle of an insect colony, more imagination is needed to appreciate the flurry of molecular communication that maintains these fungal connections with plant roots. This chemical dialogue is needed to modulate the defences of the plant and to quell tissue damage by the fungus. In some mycorrhizal symbioses it is impossible to be sure which, if either, organism benefits from the interaction. The issue is complicated by the fact that the behaviour of a fungus can sometimes shift from mutualism to parasitism. The edible matsutake mushroom, *Tricholoma matsutake*, performs this switch. Its mycelia form mycorrhizas with young roots, then turn nasty and attack the host plant, and, finally, live as saprotrophs by decomposing the dead tissues of their former collaborators.

Fungi establish many kinds of mycorrhizal relationships with plants (Figure 21). Orchid mycorrhizas illustrate the complexity of these associations. The Orchidaceae is one of the largest families of flowering plants, comprising 25,000 to 30,000 species. Orchid seeds are microscopic and do not contain sufficient nutrients to support germination and early seedling development. Nourishment for these critical stages in development is provided by fungi that grow as knots of convoluted hyphae inside the cells of the swelling embryo. These knots are digested by the orchid, furnishing the little plant with food until it turns green and can start supporting itself by photosynthesis. The advantage of this connection to the fungus is unknown and it seems plausible that it is the victim of parasitism by the orchid. As the plant grows, however, a more hospitable relationship develops, with the fungus providing the plant with

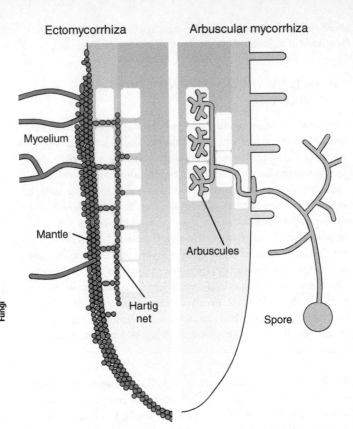

21. Two types of mycorrhizal association between fungi and plant root systems.

phosphorus and nitrogen and the orchid contributing sugars. The nutrients supplied by the fungus are absorbed from the soil by its mycelium. Mycorrhizal fungi operate as accessory root systems for plants.

The fungi in orchid mycorrhizas can connect with the roots of other kinds of plants to create 'common mycorrhizal networks' that support nutrient flow between trees, fungi, and orchids. Pale

60

orchid species that do not produce chlorophyll exploit these networks by obtaining all of their food from the continuous digestion of the knots of fungal hyphae in their cells. These orchids are called mycoheterotrophs, referring to their fungal diet, and mycorrhizal cheaters for their unusual parasitic lifestyle. Photosynthesis by the trees is the ultimate source of the food that sustains these mycoheterotrophs, with the fungus being exploited as the go-between. Other kinds of plants, including species of Ericaceae that lack chlorophyll, engage in similarly manipulative relationships with fungi.

Fungi that form mycorrhizas with orchids are basidiomycetes. Many of these species are called jelly fungi because they produce fruit bodies with a rubbery or gelatinous texture on rotting wood. Basidiomycetes with umbrella-shaped mushrooms establish a different kind of mycorrhiza with trees and shrubs. Hyphae of these 'ectomycorrhizal' fungi wrap around the tips of rootlets forming a tight glove, or mantle, and grow between cells in the outer layers of the root tissue creating a structure called the Hartig net. Dissolved minerals are delivered to root cells through the Hartig net and sugars flow to the fungus in the opposite direction. Ectomycorrhizal fungi include boletes, with pores beneath their caps, and many of the gilled mushrooms common in forests. Species of *Amanita* (including the fly agaric and death cap), *Cortinarius* (webcaps), *Laccaria*, *Lactarius* (milk-caps), and *Russula* are common ectomycorrhizal fungi with a global distribution. Ectomycorrhizas are formed with only 3 per cent of plant species, but their ecological and economic importance is enormous because these plants include conifers that dominate the northern boreal forests and rainforest trees called dipterocarps in Southeast Asia.

Broadleafed trees in temperate forests and some of the dominant tree species in African and South American forests are also supported by ectomycorrhizas. Ectomycorrhizal fungi have been overlooked in some ecosystems because they produce mushrooms infrequently. Arbuscular mycorrhizas are also essential for the health of forests and they are the commonest type of mutualism between fungi and plants. Arbuscular mycorrhizas are formed between fungi classified in the Glomeromycota and 70–90 per cent of plant species including liverworts, ferns, gymnosperms, and flowering plants. Physical contact between fungi and plants in these mergers is provided by arbuscules, which are finely branched connectors between fungal mycelia and the interior of root cells. Arbuscules provide a large surface area for nutrient exchange. Experiments have shown that these symbioses increase plant productivity by as much as 20 per cent. Arbuscular mycorrhizas have huge agricultural significance because they are formed with wheat, corn, rice, potato, and other crops.

If a mushroom is fruiting beneath a healthy tree it is tempting to assume that it is formed by a mycorrhizal fungus that is connected to the tree. DNA fingerprinting can be used to test this idea and often reveals an unexpected level of fungal diversity. In one experiment in Scotland, investigators found that individual pine trees were associated with the mycelia of up to nineteen different mushroom colonies. The mycelia of ectomycorrhizal fungi are formed by branched hyphae, and thicker assemblages of hyphae called 'cords' help fungi negotiate dry soil patches and bridge areas where nutrients are scarce. The mycelium connected to every metre of root length can thread its way through hundreds of metres of surrounding soil. This spreading growth pattern is a good tactic for finding soil patches that are rich in nutrients. The protein in the carcass of a dead insect, for example, offers a dense source of nitrogen for the mycelium. Digestion of the insect is followed by distribution of nitrogen to distant parts of the mycelium and transmission to the plant root. Mycorrhizal fungi can also access essential mineral nutrients by penetrating rock surfaces and secreting organic acids that dissolve calcium, magnesium, and other elements.

Ectomycorrhizas and arbuscular mycorrhizas create mycorrhizal networks like those associated with orchids. By linking adjacent root systems, the fungi support populations of a single plant species and communities of multiple plant species. Experiments show that plants warn one another about pest attack by passing signalling molecules through their common mycorrhizal networks. This allows plants to produce defence compounds before they are attacked by aphids or caterpillars. By operating as an underground communication system for plants, mycorrhizal networks help to protect their hosts and thereby maintain their personal supply of sugars.

The functions of individual mycorrhizas and common mycorrhizal networks in the ecology of forests, grasslands, and other habitats makes it essential for ecologists to consider the activities of fungi in models of plant productivity. The attention to plants as isolated ecological entities was one of the reasons that the study of ecology progressed so slowly and answered few fundamental questions in the 20th century. Mycorrhizas are no longer an afterthought for scientists who develop models of the flow of energy through ecosystems. Mycorrhizas are also an important consideration in the evolution of land plants. Fossils of spores produced by arbuscular mycorrhizal fungi suggest that these symbioses were formed with early land plants. The first of these relationships may have resembled the arbuscles found in today's liverworts, which are among the most 'primitive' plants. Life on land could not have arisen without mycorrhizal fungi and remains dependent on continuing mutualisms between fungi and plants.

Fungi called endophytes form a very different kind of mutualism from mycorrhizas by housing themselves inside plant tissues without any connection to an external mycelium. Endophytes live in the stems and leaves of plants and do not enter their roots. They grow in spaces between plant cells and along the walls of adjacent cells, but do not form structures that resemble the Hartig net or penetrate plant cells with anything like an arbuscle. Endophytes are ascomycetes that appear to have evolved from pathogenic fungi that infect plants and insects. The resemblance between the genomes of a fungus that destroys rice and a rice endophyte is one example of an obvious connection. *Neotyphodium* species, which are endophytes that live in grasses, are related to insect pathogens. The presence of *Neotyphodium* in a plant results in faster growth rates, greater tolerance to drought, and resistance to other fungi that cause disease. These endophytes also produce toxic alkaloids that act as natural pesticides that deter insect damage. Some of these compounds are poisonous to horses and cattle, causing the constriction of blood vessels in the animals' extremities and producing a condition called fescue lameness. These symptoms are similar to ergotism in humans and other mammals, which is caused by the fungus *Claviceps purpurea*. This is not surprising because the ergot fungus is another close relative of endophytic species of *Neotyphodium*.

Neotyphodium has lost the ability to reproduce sexually and its hyphae are dispersed inside the seeds of its host plants. This 'vertical transmission' allows the fungus to move directly between generations of plants. (In medicine, vertical transmission is used to describe a bacterial or viral infection transferred by a pregnant mother to her foetus.) Related endophytes produce sexual ascospores as well as asexual conidia and use these to sweep across a population of grasses. This form of airborne transfer is referred to as horizontal transmission, and is comparable to the way that a cold virus is spread in an office. Horizontal transmission is very common among endophytes that colonize trees and woody shrubs. Endophytes that protect plants from diseases caused by other microorganisms can switch to the decomposition of leaf tissues towards the end of a growing season. Other endophytes living in the sapwood of trees can begin the active breakdown of wood as its host ages or is damaged. These are additional instances of the plasticity of associations between fungi and plants that allow mutualistic fungi to become parasites.

Lichens

Lichens are the best-known mutualisms involving fungi. They are composite organisms consisting of a fungus and a single-celled alga or cyanobacterium. Algal and cyanobacterial partners in lichens provide the fungus with food produced by photosynthesis. Swiss biologist Simon Schwendener made this discovery in the 19th century. The nature of lichens as coalitions between different organisms was a controversial idea at a time when scientists tended to regard species as isolated entities whose interactions were limited to eating and being eaten. Beatrix Potter, the famous author of children's books, became interested in lichens and was one of the people who resisted Schwendener's proposal. (Potter conducted experiments on spore germination, but her modest contributions to mycology have been hyped to an absurd degree when the work of so many other women scientists deserves greater recognition.)

Despite this opposition, many eminent biologists of this period recognized that lichens were 'dual organisms' and discovered other mutualistic relationships including ectomycorrhizas, root nodules (bacteria and legumes), and corals (algae and animals). Scientists that specialize in the study of lichens are called lichenologists and they tend to work alongside bryologists who study bryophytes (mosses, liverworts, and hornworts). This intellectual alliance makes sense because lichens and bryophytes often grow in the same habitats, which means that experts on one group of organisms are always finding species of the other group. In the polar tundra, for example, where there are few vascular plants, miniature forests of lichens and bryophytes are the primary producers. Arctic lichens serve as important winter forage for caribou (reindeer).

The Latin names of lichens refer to the species of fungus rather than the photosynthetic companion. This makes a lot of sense because lichens are formed by 18,000 species of ascomycetes (40 per cent of all ascomycetes) with only 150 species of green algae and bacteria. While the fungi in lichens cannot grow without their photosynthetic symbionts, the green algae and bacteria certainly thrive outside lichens. They are recruited from the environment by the fungi and become cradled within a tough fabric of interwoven hyphae in the body, or thallus, of the lichen. This 'fungus first' mechanism explains how a single lichen thallus contains one fungus but can incorporate multiple strains of the photosynthetic partner.

Lichens are categorized by shape as crustose (thin crusts), fruticose (shrubby), and foliose (leafy). *Rhizocarpon geographicum* is a crustose lichen that grows on rocks in the form of a mosaic of yellow patches separated by cracks. With a little imagination, the yellow patches look like regions on a map with the cracks serving as boundaries. This explains the common name of yellow map lichen. In the temperate rainforests of the Oregon Cascades, long grey-green strands of the Methuselah's beard lichen, *Dolichousnea longissima*, hang from the branches of Douglas firs. Growth of this fruticose lichen is very sensitive to atmospheric pollution and it has disappeared from much of its original range in Europe. Other lichens flourish under unpromising conditions of industrial and agricultural pollution. *Xanthoria parietina* is an orange or yellow foliose lichen that grows on roofing in urban areas and covers farm buildings where it is exposed to nitrogen fertilizers. Its natural preference for high nitrogen levels is also shown in its appearance in coastal areas on rocks fertilized by bird guano. Variations in the sensitivity of lichens to nitrates, sulphates, and other chemicals make them very useful as indicators of industrial and agricultural pollution.

Most lichens contain green algae, but 15 per cent of species involve cyanobacteria. In addition to the formation of sugars by photosynthesis, the cyanobacteria in lichens absorb nitrogen gas (N_2) from the atmosphere to produce ammonia (NH_3). The ammonia is used by the bacterium and the fungus as a source of nitrogen for the synthesis of proteins and nucleic acids. This is crucial for lichens living in places where nitrogen is scarce. Species of *Peltigera*, called dog or pelt lichens, form large grey, green, or brown lobes on wet soil and other mossy locations. Some species of these foliose lichens are three-way combinations of fungus, green alga, and cyanobacterium. Fungi absorb sugars and exchange other nutrients with 'their' algae through slender branches that protrude into the green cells. Contact with cyanobacteria is achieved by forming tight connections with the cell surface without penetrating the bacterial wall.

Lichens reproduce by releasing spores of the fungus and rely upon young colonies of the fungus to associate with their photosynthetic partners soon after spore germination. This is only effective, of course, when the appropriate algae or cyanobacteria are prevalent in the environment. Ascospores are produced in fruit bodies that develop on the surface of lichens. The red-tipped branches of a lichen called British soldiers, *Cladonia cristatella*, are fruit bodies displaying a layer of asci and intervening cells that contain globules of red pigment. The common name of this lichen refers to the 'Redcoats' who fought against the Continental Army during the American Revolutionary War. A minority of lichens are produced by basidiomycetes rather than ascomycetes. These lichens reproduce by forming tiny mushrooms. In addition to the release of sexual ascospores and basidiospores, lichens reproduce asexually by dispersing small parcels of fungal hyphae wrapped around algal cells. These form as a powdery coating on the thallus and reformulate exactly the same symbiosis in a new location.

Some of the mutualistic relationships examined in this chapter are bewilderingly complex. Mushroom farming by ants and termites seems particularly intricate because it is dependent on so many physical and molecular interactions between the participants. But we misread biology by treating symbiosis as an exceptional phenomenon. No species exists without benefit from interactions with many other organisms. Fungi, along with other microbes, associate with every animal and plant. Many of these relationships are not as obvious as the growth of algae inside lichens or the development of sheaths of mycorrhizal fungi around root tips, but they are every bit as intimate. Insect guts may seem an unlikely place for complicated symbioses, but they teem with fungi. Researchers have identified hundreds of new yeast species inside beetles and know very little about what these fungi are doing. Some of them may be commensals, benefiting from food in the host digestive system without causing any negative effects, but mutualistic interactions are probably very common too. Detrimental relationships between pathogenic fungi and plants are similarly diverse and these are the subject of Chapter 5.

Chapter 5
Fungi as parasites of plants

Global impact

Parasitic fungi that grow on plants have reshaped the biosphere and caused the deaths of millions of people since the beginning of agriculture. Dutch elm disease and chestnut blight are examples of fungal pandemics that resulted in widespread ecological changes in the 20th century. These diseases destroyed billions of trees, remodelled the urban landscape in Europe and North America, and ravaged the broadleaf forests of the Eastern United States. Recent pandemics include ash dieback in Europe, and needle blights of pines in Europe and North America. The appearance of these tree diseases is a consequence of global commerce and the unwitting introduction of fungi that can attack plant species with no resistance to foreign pathogens. Climate change may be a contributing factor in the vulnerability of trees to fungal infection and is a major concern for forest management. The sight of dying trees is disturbing, but the wider ecological consequences attract little lasting attention. This is the way with the continuous remodelling of forests and other ecosystems by disease and human activity. People born after a destructive phase in forest history overlook the changes because they have no experience of the earlier richer environment.

Fungi that infect crops have more immediate and lasting effects on human populations. Crop failures have caused starvation, economic collapse, social conflict, warfare, and mass emigration. Rusts and smuts that attack cereals have been a source of immense misery throughout history and comparable suffering has resulted from other kinds of fungi that destroy rice. As the human population swells, scientists are engaged in a global initiative to understand and combat fungi that infect staple crops. Farmers and consumers are affected by fungi that damage fruit crops including cacao, citrus, coffee, and grapes, and a wider range of commodities including natural rubber. The study of fungal diseases of domesticated and wild plants is part of the wider field of plant pathology that also includes research on viruses, bacteria, protists, nematodes, and insects.

Rusts

Stripe rust is a highly destructive disease of wheat that poses a significant threat to global food security. It is widespread across wheat-growing regions and has been called the 'polio of agriculture'. Stripe rust, also known as yellow rust, is caused by the rust fungus *Puccinia striiformis*. This microorganism feeds on wheat and a thorny shrub called barberry, alternating between these plant hosts according to its stage of development (Figure 22).

Stripe rust on wheat produces lines of yellow blisters between the leaf veins that are filled with spores. The spores in the blisters are urediniospores (also called urediospores and uredospores) and spread the infection through a crop when they are dispersed by wind. When urediniospores germinate on wheat, their filamentous hyphae extend over the leaf surface and are capable of detecting tiny ripples associated with the underlying cells. The sensitivity of the fungus allows it to angle its growth across the leaf rather than along the length of the blade. By growing in this direction, the fungus maximizes the likelihood of finding open breathing pores, or stomata, which offer easy access to the interior of the plant. Germination of the spores is timed to coincide with the cooler morning temperatures when the stomata are open. Wheat plants close their stomata in late morning to reduce water loss.

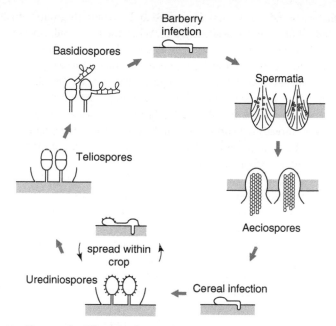

Barberry infection

Basidiospores

Spermatia

Teliospores

Aeciospores

spread within crop

Urediniospores

Cereal infection

22. **The complex life cycle of a cereal rust.**

As the cereal infection progresses, the rust begins to produce a second type of spore called the teliospore in the same blisters. Teliospores are dark brown. These spores do not infect wheat. They serve instead as a reservoir for the disease, surviving in the dead leaves or the soil. When environmental conditions are suitable, the teliospores germinate and produce a third type of spore. These are basidiospores. This developmental cycle is very complicated, but the challenge of writing and reading this step-by-step description is necessary for an understanding of the biology of the rust fungi. On a positive note, comprehension of the rust life cycle makes everything else in fungal biology seem quite straightforward.

Rust basidiospores are formed by the same process that produces spores on the surface of mushroom gills. This similarity in development is not surprising because genetic evidence shows that rusts and mushrooms are related and mycologists group both kinds of fungi in the basidiomycetes. When the rust basidiospores are dispersed in air they infect the alternate host, the barberry shrub. Rather than searching for open stomata on barberry, the rust punches its way straight through the leaf surface. This direct penetration process is widespread among fungi that infect plants. The subsequent damage done to barberry does not concern cereal growers directly. The problem with the shrub is that the fungus uses its leaves as a platform for producing a fourth type of spore, called the aeciospore, which infects wheat.

The formation of aeciospores occurs after two strains of the rust unite on the barberry leaves. Rather than growing towards one another and fusing like mushroom colonies, however, cells of the rust are carried by insects that fly from shrub to shrub. The rusts produce sugary nectar that attracts flies and the cells, called spermatia (or pycniospores), stick to the body of the insects as they feed. This is comparable to the process of insect pollination in flowers. After 'pollination', aeciospores are formed and are shed from the underside of the barberry leaves and infect wheat leaves. Each aeciospore carries a pair of nuclei, one from each of the strains brought together by the insects. This means that the aeciospores are heterokaryotic, like the mycelia of mushrooms whose compartments contain pairs of nuclei after two homokaryons fuse in the soil.

In a more remarkable instance of fungal pollination, the rust *Puccinia monoica* prevents flowering in mustard plants and turns the green stem and leaves of its host into bright yellow imitation flowers. Bees and other insects are attracted by the yellow coloration and aromatic molecules released from the infected tissues. As they feed on nectar associated with the fungus, they become covered with its spermatia and carry them to other 'flowers' where the pollination process is completed. After pollination the rust produces airborne aeciospores that infect grasses, which are its alternate hosts.

The formation of urediniospores, teliospores, basidiospores, spermatia, and aeciospores may seem a fantastically convoluted way of life, but each phase of development serves a critical function in perpetuating the rust fungus. Rust pollination creates new combinations of nuclei from compatible mating types of the rust and these are transmitted through the urediniospores and teliospores. Fusion of nuclei and meiosis in the teliospores produces basidiospores whose single nuclei contain novel combinations of the genes from the two mating types. Again, these events are comparable with the formation of homokaryons, heterokaryons, and spores in mushrooms described in Chapter 3.

Intensive agriculture has allowed cereal rusts to sidestep the barberry part of their life cycles. With the cultivation of summer and winter wheat, the fungus has a source of food for most of the year and can attack successive crops using urediniospores alone. Warmer temperatures associated with climate change may be exacerbating this problem. In regions where winter temperatures are low enough to kill the urediniospores, the barberry stage can still be omitted by the spread of spores from infected wheat in warmer areas. Over many growing seasons, the edited life cycle may weaken the fungus by preventing the genetic recombination that takes place on barberry. The high levels of genetic variation among rusts suggest that they avoid this handicap by maintaining a background of barberry infection even when their urediniospores spread over long distances. This intermittent growth on barberry allows the fungus to adapt quickly to the introduction of new disease-resistant varieties of wheat developed through crop breeding programmes.

Airborne urediniospores spread stripe rust over thousands of kilometres in a stepwise fashion from crop to crop of susceptible plants. In North America, for example, cereal rusts spread northward along an established 'Puccinia pathway' that runs from Mexico to Canada. In a stripe rust epidemic in the 1950s, for example, the disease spread 2,400 kilometres from northern

Mexico to North Dakota in six months. Movement of spores according to weather conditions and the availability of susceptible crops allows cereal rusts to spread in other wheat-growing regions. In China, for example, stripe rust disperses across wheat-growing provinces on easterly winds, creating a wave of epidemic disease over hundreds of kilometres of farmland. Rare cases of disease introduction also occur via a single-step invasion when a fungus moves over a great distance without infecting crops on the way. Stripe rust was carried from Europe to Australia on infected plants, or possibly on clothing, in the 1970s and then spread to New Zealand by windblown spores in 1980.

Black stem rust, caused by *Puccinia graminis*, was the probable culprit for the destruction of crops in the Roman Empire of the 1st century AD. These epidemics occurred during a period of wet and cooler weather and the resulting food shortages played a part in the decline of Roman civilization. Disease resistant varieties of wheat have been effective at reducing the impact of this disease, but the fungus continues to cause crop losses. A highly virulent strain of black stem rust discovered in Uganda in 1999 has spread to other wheat-growing countries in Africa and to Yemen and Iran. The evolution of new virulent strains of this rust is a relentless process. Our reliance on monocultures of cereals, and other crops, means that a single race of a fungus that can overcome the defences of its host has the potential to destroy a commodity that is grown over a wide geographical area.

Rusts are examples of biotrophic pathogens that feed on living plant cells without killing the host. Biotrophy enables the fungus to withdraw nutrients for days or weeks while it keeps producing spores at a feverish rate. The dependence of rusts on living host tissues is so unwavering that they cannot be grown in the laboratory on culture plates. Necrotrophs kill plant cells and feed on their contents. Necrotrophic growth is responsible for chestnut blight and Dutch elm disease (both caused by ascomycete fungi). Many fungi confuse this simple categorization by starting with a biotrophic interaction and shifting to necrotrophic behaviour as the disease progresses.

Parasite and pathogen are catch-all terms that many mycologists apply to biotrophs and necrotrophs. Rusts form colonies of branching hyphae that grow between leaf cells and absorb nutrients through structures called haustoria that protrude into the cells. Haustoria are similar to the arbuscules produced by mycorrhizal fungi because they dimple the membrane of the plant cell without breaking it. The resulting interface is like a hand (the haustorium) inside a glove (the membrane of the plant cell) and the fungus absorbs food across this tight connection.

Plants mount an immune response to attack by rusts. The plant is furnished with receptors that induce a cascade of defences when they are activated by molecules released by the fungus. There are two defence systems: a non-specific response to a common family of chemicals released by all microbes, and a specific response to fungi recognized by the plant. These are analogous to the innate and adaptive immune systems that have evolved in animals. One of the specific defence reactions is called the hypersensitive response. This triggers the death of plant cells in the immediate vicinity of the early infection, which has the effect of creating a barrier of dead cells that obstructs the spread of the fungus. In a continuing arms race between pathogens and hosts, an array of molecular processes has evolved in the rusts to thwart the plant defences.

More than 7,000 species of rusts have been described with a host range that extends from cereals to other flowering plants, conifers, and ferns. Many rusts do not engage in the complex life cycle described for the cereal rusts and only infect one plant species. The fungi that cause Asian soybean rust (*Phakopsora pachyrhizi*) and coffee rust (*Hemileia vastatrix*) are examples of rusts with this kind of truncated life cycle. They infect the leaves of their hosts with urediniospores and do not seem to engage in the pollination process and aeciospore formation on another plant. This is strange, because both of these rusts form teliospores and shed basidiospores into the air. It is possible that the basidiospores infect another plant species and that this cryptic phase of the life cycle has not been found. Logical places to search for these missing hosts would be East Asia, for soybean rust, and Ethiopia, for coffee rust, where the target crops originated.

Smuts

The smuts are another group of basidiomycetes that cause plant disease. Most of the 1,400 species of smuts infect grasses, including cereals, and sedges. Each smut is dedicated to killing a single plant species or handful of related plants and does not switch to an alternate host to complete its life cycle. Smuts do not produce as many spore types as the rusts. Sugar cane smut, caused by *Sporisorium scitamineum*, occurs in all areas where the crop is cultivated. The use of sugar cane in the production of biofuel ethanol has increased the interest in this pathogen. Infected plants bear a blackened stalk called a smut whip that is filled with teliospores. These spores can be spread to other plants or deposited in the soil.

When the smut teliospores germinate they form a short outgrowth from which a second spore type is produced. These spores are similar to the basidiospores of rusts, but they are called 'sporidia' by experts on smuts. Sporidia bud to form yeast cells and fusion of two of these yeast cells from different strains creates the invasive form of the fungus that penetrates the sugar cane plant. The smut induces the premature formation of flowering tassels by upsetting the normal hormonal balance in the plant. The normal tassel is a tall feathery structure called an arrow, but the fungus transforms this into the smut whip.

Similar growth processes play out in head smut of corn and covered smut of sorghum caused by other species of *Sporisorium*.

The related smut, *Ustilago maydis*, infects the ovaries of corn and converts the kernels into swollen bags of teliospores. Infected corn kernels are used as a flavourful ingredient called huitlacoche in Mexican cooking. Huitlacoche was part of Aztec, Hopi, and Zuni cuisine long before the Spanish conquests in the 16th century. The yeast phase of the smut grows in culture, which makes it easier to manipulate than rusts that cannot be grown separately from their plant hosts. *Ustilago maydis* has been used as a model for cancer research. Disruption of a gene in the smut fungus called *brh2*, which is related to the human tumour suppressor gene *BRCA2*, results in a deficiency in DNA repair mechanisms. This is consistent with a mechanistic link between mutations in the human gene and an increased risk of developing breast cancer.

Some smuts show astonishing finesse in the way that they control the development of their hosts. A beautiful example of this parasitic manipulation is seen in the infection of campion flowers by the smut *Microbotryum violaceum*. When the fungus infects the female flowers it suppresses the formation of the ovaries and stimulates the production of the male organs called stamens. Normal stamens bear anthers that hold pollen grains, but the fungus replaces the pollen with its teliospores. This subversion of floral development allows the smut to make use of butterflies and other insect pollinators to disperse its spores.

Ascomycetes and other plant pathogens

Trees are plagued by a staggering number of fungi. Chestnut blight and Dutch elm disease are caused by ascomycetes that destroy the vascular tissues that convey water and nutrients through trees. *Cryphonectria parasitica* is the chestnut blight pathogen, whose spores are spread by wind, insects, and birds. Bark beetles that chew tunnels in the sapwood of elm trees transmit the spores of *Ophiostoma ulmi* and *Ophiostoma novo-ulmi*, responsible for Dutch elm disease. Ash dieback, which was first identified in Britain in 2012, is caused by windblown spores of *Hymenoscyphus pseudoalbidus* (also known as *Chalara fraxinea*). Symptoms include leaf loss and dieback of branches and twigs. Needle blights of conifers are produced by ascomycetes that do not spread beyond the needles. They do not kill trees outright, but debilitate the host slowly by damaging the foliage year after year. Ascomycetes that produce anthracnose diseases of broadleafed trees have the same effect.

Powdery mildews are ascomycetes that attack plant species in many different families, and damage crops including wheat and barley, grapes, onions, apples and pears, cucumbers, and strawberries. The common name refers to the formation of chains of spores that coat infected leaves in a white powder. Powdery mildews produce haustoria in the outermost cells of the leaf and develop an extensive mycelium on the surface of the host. Most species do not penetrate deeper into the plant tissues. But despite the superficial nature of the invasion, powdery mildews cripple their hosts by diverting nutrients to support the mycelium. They suck the plants dry. The haustoria of some mildews are simple knobs, but *Blumeria graminis*, that infects barley and other grasses, forms haustoria with long finger-like extensions.

Powdery mildews are biotrophs like rusts and cannot be grown in culture. Compared with the enormous genomes of rusts (described in Chapter 3), *Blumeria graminis* functions with far fewer protein-encoding genes. The reason for this difference is not clear, but the loss of genes is characteristic of many parasites that establish obligatory relationships with single hosts. One reason that rust genomes are so big is that these fungi need to overcome the defences of two unrelated plant species. The formation of so many types of spore is another characteristic that may be related to genome expansion.

The chains of spores produced by powdery mildews are asexual conidia. Mycelia of compatible mildew strains associate on the surface of leaves and produce sexual ascospores inside pinhead-sized fruit bodies. The fruit bodies are decorated with long hairs that bend as they dry. In some powdery mildew, these appendages act as stilts that lever the fruit body above the tangled mycelium on the leaf surface. This allows the sexual stage to fall free from the leaf and overwinter on surrounding twigs. In the spring, the fruit bodies crack open along a line of weakness, exposing the asci within. Explosive discharge of the asci blasts the ascospores into the air where they are dispersed by wind.

Mildew ascospores are infectious and can establish a new round of disease in a crop. In grape powdery mildew, for example, the fruit bodies develop at the end of the summer in wine-growing regions and remain dormant during the winter. Ascospores are released when the fruit bodies open in the spring, and these initiate infections when they land on young leaves. As the disease develops, the fungus colonizes shoots and ripening berries and its asexual conidia spread the mildew throughout the vineyard. Powdery mildews are among the enchanting illustrations published by Charles and Louis-René Tulasne in 1861 (Figure 23). The Tulasne brothers discovered the way that single species of ascomycetes rotate between the formation of asexual and sexual spores (Chapter 2).

Powdery mildew is the most damaging disease of grapes, but other ascomycetes produce black rot, bitter rot, and ripe rot in vineyards. Adding to the challenges posed by multiple pathogens, a single fungus is capable of producing several diseases depending on the timing of the infection. *Botrytis cinerea* is an ascomycete that grows as a necrotroph and produces bunch rot when it develops on fruit damaged by insects or high winds, and grey rot in wet weather. The same fungus also causes noble rot when its growth is stimulated by rainfall, followed by weeks of hot, dry weather. Noble rot results in the drying or partial raisining of the grapes, which concentrates their flavour and sweetness. Grapes with noble rot are picked individually to produce expensive dessert wines like Sauternes from Bordeaux.

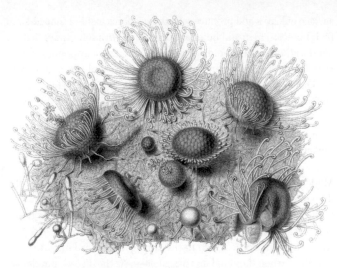

23. The powdery mildew fungus *Microsphaera penicillata*, illustrated by Charles Tulasne.

In rice blast disease, caused by the ascomycete *Magnaporthe grisea*, conidia stick to the leaf surface, germinate to produce a short hypha, and form a swelling at the end that serves as a pad from which the infection proceeds. This pad is called an appressorium (Figure 24). It is a rounded cell that fastens itself to the rice leaf with an adhesive O-ring and blackens itself with melanin pigment. The pigment prevents leakage of molecules from the cell as it swells and becomes pressurized. This pressure comes from the absorption of water and rises to ten times the level that inflates a bicycle tyre. Once pressurized, the fungus produces a thin peg on the underside of the cell that pierces the plant surface. In some diseases, fungi release enzymes to weaken the plant surface to allow easier access. The rice blast fungus is sufficiently strong to work by pressure alone. It deforms metal films in laboratory tests and pushes holes through plastics including the bulletproof vest material called Kevlar.

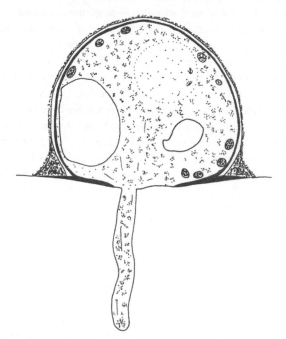

24. Appressorium of the rice blast fungus, *Magnaporthe grisea*.

Rice blast can be controlled with fungicides, but recent epidemics have ruined millions of hectares of crops. The devastating effects of rice blast fungus have led to concerns that it might be used as a biological warfare agent. Before the United States halted research on bioweapons during the Nixon administration, field trials were conducted in the 1960s to gauge the potential impact of the fungus on rice production in China and SouthEast Asia.

Rice is attacked by many other ascomycetes at the seedling stage and later in plant development. Infection by *Fusarium fujikuroi* causes plants to elongate, topple over, and die without producing any grain. These unusual disease symptoms are induced by the production of a growth hormone called gibberellic acid. Plants synthesize this hormone to stimulate seed germination and control their growth rate, but the production of an excess of the compound by the fungus has disastrous effects on rice. One thousand species of *Fusarium* have been described and many of them are plant pathogens. Banana growers are very concerned by a virulent strain of *Fusarium oxysporum* called Tropical Race 4. The global banana crop is exceedingly vulnerable because the Cavendish cultivar, which has served as the commercial banana since the 1960s, is a clone derived from a single plant grown in Asia. The previous cultivar of banana was wiped out by Fusarium wilt, also known as Panama disease, in the 1950s, and Cavendish has worked well since then. But with the emergence of the TR4 strain of the fungus the entire crop is threatened again.

Rusts, smuts, and ascomycetes are responsible for most fungal diseases of plants. A handful of zygomycetes cause soft rots of flowers, fruits, and bulbs. The zoosporic fungus *Olpidium brassicae*, described in Chapter 2, infects cabbage roots, but, like zygomycetes, these basal groups of fungi are more important in decomposition than plant disease. It is worth revisiting the fact that many plant diseases are due to microorganisms that look like fungi but are not fungi. For example, the potato blight pathogen, *Phytophthora infestans*, is an oomycete water mould (Chapter 1). Pathogens called downy mildews are also oomycetes. Grapes are infected by the ascomycete powdery mildew *Uncinula necator*, and the oomycete downy mildew *Plasmopara viticola*. A practical consequence of this taxonomic nuance is that some fungicides that affect powdery mildews do not work against downy mildews.

Fungicides

Bordeaux mixture, which contains copper sulphate and slaked lime (calcium hydroxide), was developed as a preventative spray against grape diseases in the 1880s. The effectiveness of copper as a fungicide was probably recognized much earlier than this and it is still used today. Bordeaux mixture prevents spore germination on the leaf surface and can also be applied to soil to control root disease. Fungicides range from blends of simple compounds containing copper and sulphur, to generations of synthetic compounds that disrupt specific metabolic pathways in the fungi. Contact fungicides are active on plant surfaces and systemic fungicides are absorbed by the plant and transmitted throughout its tissues. Mancozeb is used as a protectant in the same way as Bordeaux mixture. This broad-spectrum fungicide interferes with lipid synthesis and energy production in fungi. Systemic fungicides include carboxin that inhibits protein synthesis, and triadimefon that is applied as a seed treatment and damages the cell membranes of fungi.

The development of resistance to fungicides has affected the usefulness of many compounds within a few years of their introduction. Benomyl, which disrupts the protein skeleton inside fungal cells, was used against fungal diseases from the late 1960s until its withdrawal from the market in 2001. It was a very effective treatment for powdery mildews and a wide range of other diseases, but resistance spread very swiftly and there were concerns that human exposure to benomyl was associated with birth defects. The limited 'shelf life' of fungicides, coupled with their expense and potential toxicity towards humans and other animals, explains why there is so much investment in breeding programmes to exploit natural defence mechanisms of plants against fungi.

Chapter 6
Fungi and decomposition

Fungi and the carbon cycle

Most people associate fungi with rotting fruit and decaying wood, and, more generally, regard their presence as a symptom of death. This is not unreasonable. Fungi that do not form supportive or parasitic relationships with plants and animals feed on the debris of life. These saprotrophs decompose dead roots, leaves, flowers, fruits, seeds, twigs, branches, upright tree trunks, and fallen logs. They rot animal faeces and the tissues of invertebrates and vertebrates. Some of these decomposers produce mushrooms and these become a source of nutrients for other organisms. Fungi also clear up our mess, breaking down every natural product used in a lifetime of consumerism and destroying synthetic materials made by industry. Fungi spin the carbon cycle.

Coprophilous fungi

Cows and other ruminants have a complex digestive system that is adapted for processing fibrous plant materials. After chewing and rumination (rechewing), forage is digested by a combination of enzymes secreted by the cow and by microorganisms inside the animal's stomach compartments and its gut. Anaerobic fungi participate in cellulose breakdown in the rumen. When the dung is dropped on the ground, the immediate availability of oxygen stimulates the growth of 'coprophilous' fungi that permeate the dung and absorb nutrients from the residue of plant materials.

Cow dung does not seem like a promising place to look for beautiful fungi, but inspection of its malodorous surface with a hand lens reveals some stunning organisms. After deposition, the dung of grass-fed animals becomes covered with *Pilobolus* stalks that sparkle with fluid droplets and resemble cut crystal glassware. These pressurized structures, described in Chapter 1, function as squirt guns that blast their spore-filled capsules away from the dung. The cup fungus *Ascobolus stercorarius* is another brief occupant. The Latin name of this species means 'of or relating to dung'. This fungus fashions five-millimetre-wide yellow discs that support a layer of thousands of explosive asci. Each disc is a fruit body covered with conspicuous dots and each dot is a single ascus coloured by its violet ascospores. When spores are released by a group of asci in one part of the disc, the disturbance causes nearby asci to detonate and these mini-explosions surge from one side of the little fruit body to the other. Discharge of spores from individual asci happens too quickly to be visible without recording the mechanism with a high-speed camera running at more than 100,000 frames per second. But the salvo of hundreds of exploding asci is apparent as a wave of disappearing dots. The resulting puff of ascospores is visible as smoke rising from the fruit body.

Tiny gilled mushrooms also grow from the dung. These include the ink cap species *Coprinopsis cinerea*, and a mottlegill called *Panaeolus semiovatus*. Some hallucinogenic mushrooms are associated with animal dung. The liberty cap, *Psilocybe semilanceata*, grows in pastures fertilized by sheep, horses, and cows, but does not fruit directly on the dung.

Rather than reproducing at the same time, different fungi produce their spores according to a programme that plays out over several days. Zygomycetes, including *Pilobolus*, are the first fungi to emerge from cow dung, followed by cup fungi and other ascomycetes, and mushrooms appear last. This ecological succession illustrates the complexity of the decomposition process. The spores of *Pilobolus* pass through the digestive system of an animal before they are deposited in its fresh dung. They germinate immediately, forming an extensive mycelium within the dung that absorbs relatively simple nutrients before the development of spores. The cup fungi may also pass through a herbivore and begin growth immediately, but they invest more time in breaking down more complicated molecules in the dung before fruiting. Mushroom colonies take the longest to digest the most resilient polymers in the dung and generate enough mass to support the development of their gilled fruit bodies.

Sporormiella species grow on dung from many different animals. Historical accumulations of the spores of this ascomycete in ancient sediments have been linked to changes in the abundance of herbivores. Palynological records from Queensland, Australia, show an abrupt drop in the deposition of *Sporormiella* spores 41,000 years ago. The regional climate was stable at this time and human hunters are identified as the cause of the collapse of animal populations and consequent disappearance of their dung and the fungus. A similarly precipitous decline in spore numbers in sediment samples from lakes and wetlands in North America is found towards the end of the Pleistocene (beginning 15,000 years ago). This was caused by the eradication of the woolly mammoth, mastodon, and rhinoceros by Palaeo-Indians at the end of the last ice age. Changes in the concentrations of *Sporormiella* ascospores have also been linked to the disappearance of Madagascan megaherbivores 1,700 years ago and the 17th-century extinction of giant flightless birds called moa in New Zealand.

Some of the coprophilous fungi whose dispersal mechanisms are geared towards passage through animal guts are also found on plant products that have been processed for gardening including wood chips and wood mulch. They are opportunists that have taken advantage of these refined sources of cellulose. Woody garden supplies also attract a few of the saprotrophic fungi that decompose the tissues of dead and dying plants. Plant decomposition involves a succession of fungi and is affected by moisture levels, temperature, and other environmental variables. Broad patterns of decomposition can be determined from the microscopic study of fungi growing on rotting vegetation. Species that form their spores on the plant surface can be identified directly and other fungi can be isolated from the decomposing tissues and grown in culture. The usefulness of these experimental approaches is limited, however, because they tend to exaggerate the importance of fungi with conspicuous fruit bodies and species that grow well in culture. More recent research in which fungi are identified from DNA extracted from leaves and other samples of rotting vegetation have revealed a more detailed picture of the decomposition process.

Leaf decay

Leaf decomposition is one of the processes examined using molecular methods. Deciduous leaves are primed for breakdown by endophytes that grow inside them before leaf fall as well as by fungi that populate the leaf surface. These internal and external leaf residents are yeasts and species of filamentous ascomycetes. They do not damage the leaves until they are shed, but begin consuming sugar molecules and other simple nutrients as soon as the host defences are shut down. They provide another example of the changes in fungal behaviour that can accompany an alteration in the growth conditions offered by plants (Chapter 4). After this early surge in fungal activity, decomposition proceeds more slowly as different ascomycetes colonize the leaves and digest cellulose and other complex polysaccharides. Basidiomycetes tend to be latecomers in the succession of fungal communities on decaying leaves. They are the most effective in destroying lignin, which is the complex organic molecule that strengthens plant cell walls.

Lignin is linked to cellulose and other compounds in leaves and these polysaccharides are exposed for decomposition once the lignin is dissolved. This process may uncover cellulose for ascomycetes that cannot remove lignin for themselves, allowing them to grow in the leaf litter along with the basidiomycetes. Metagenomic methods have revealed the DNA sequences of hundreds of different fungi involved in the breakdown of leaves of single species. DNA sequencing methods can also be used to examine the variety of fungal genes encoding enzymes in the leaf samples. This is useful because variations among these genes can be linked to specific groups of fungi, providing a snapshot of the fungi involved in the breakdown of different leaf components.

The remarkable complexity of the decay process is evident from DNA sequences amplified from decomposing conifer needles. Breakdown of these stiff leaves can take several years and deep layers of needles accumulate in evergreen forests. Endophytes act as pioneer decomposers in the first year. These fungi give way to new communities of saprotrophs that arrive and digest the more resistant components of the needles. After four or five years the shape of individual leaves is no longer recognizable in the deeper layers of the leaf litter and humus is produced at the end of the decomposition process. Humus resists further breakdown for hundreds of years and is an essential component of soil structure that controls moisture content and nutrient retention.

Endophytes are also crucial in the decomposition of leaves that fall into freshwater streams. Some of these fungi are ascomycetes called 'Ingoldians', after C. T. Ingold, who discovered them in the 1930s. Ingold was an influential British mycologist whose studies on fungi spanned more than seventy years. His fungi produce exquisite spores shaped as stars, crescents, sigmoids, commas, and miniature cloves (Figure 25). These develop on the surface of leaves, float downstream, and attach to submerged leaf fragments and twigs. Ingoldians, along with other kinds of fungi, bacteria,

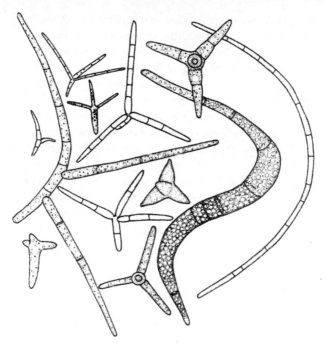

25. Aquatic spores produced by Ingoldian fungi.

and invertebrates break down leaves very swiftly in unpolluted streams. The extended shapes of the Ingoldian spores favour their attachment to leaves and they also become concentrated in foam bubbles that accumulate around rocks and fallen logs. Spores trapped in the foam may become airborne as the bubbles collapse. This would explain how these aquatic fungi establish themselves as endophytes in plants growing above the water.

Wood decomposition

Fungal decomposition of wood is more noticeable than the activities of microscopic fungi on herbivore dung and leaves. Logs in forests become decorated with the brackets of basidiomycetes that digest the masses of cellulose and lignin in the rotting heartwood. At the same time, the remaining bark becomes covered with black blobs and solid crusts of saprotrophic ascomycetes. The biochemistry of wood decay involves many different enzymes that act on cellulose, other polysaccharides called hemicelluloses, and lignin. This mixture of polymers that form the dry matter of plants is called lignocellulose. Lignin is a huge, branching molecule of aromatic rings linked into a rigid three-dimensional framework that is very resistant to breakdown. Its decomposition is carried out by fungi that secrete enzymes called peroxidases and laccases that weaken different parts of the structure of lignin. White rot fungi use these enzymes to oxidize the lignin in fallen trees and then feed on the cellulose exposed by removal of the lignin. The loss of lignin accounts for the bleaching of the rotten wood. White rot fungi include *Phanerochaete chrysosporium*, whose fruit bodies grow as thin crusts on the wood surface, and *Trametes versicolor*, known as 'turkey tail' for the concentric patterns of coloured stripes on its little brackets. The cultivated shiitake mushroom, *Lentinula edodes*, is another white rot fungus. Its capacity for growth on hardwood logs is the basis for a global industry with an annual production of more than two million tons of the flavourful mushroom.

Ganoderma applanatum is a white rot fungus with very striking fruit bodies. Its brackets have a brown top and smooth white underside and extend horizontally from trees like bookshelves. The lower surface is perforated with pores that are the open ends of spore-producing tubes. The pores are barely visible without a hand lens. A new layer of tubes grows on the bottom of the bracket each year and some fruit bodies live for years until the tree trunk is completely digested. *Ganoderma* brackets that are the size of a laptop computer release five trillion (5×10^{12}) basidiospores in a growing season from an array of two million tubes. If you crouch below one of these fruit bodies and move your head until the sun casts a beam of light just beneath the white tubes, waves of sparkling spores—fairy dust—can be seen cascading from the bracket. The common name for this fungus is artist's conk, which refers to the etchings that can be made on the white pore surface using a stylus. In the author's opinion, collection of this giant fungus to serve the whims of woodland artists should be discouraged. The live fungus on its log is beautiful enough.

The largest recorded fruit body is produced by another white rot fungus called *Phellinus ellipsoideus*. This was discovered in 2010 on the underside of an oak log in old-growth tropical woodland on Hainan Island in China. During twenty years of growth, this fungus formed a metre-long crust that weighed 500 kilograms and shed an estimated one trillion spores per day from hundreds of millions of pores.

White rot fungi include a family of ascomycetes called the Xylariaceae. *Daldinia concentrica* is a common species in the Xylariaceae. Its fruit bodies are hard black balls that expand each year, like *Ganoderma*, producing a new layer of tissue on the surface. Young specimens are pea-sized; older ones are the size of walnuts. Common names for this fungus include King Alfred's cakes, referring to the legendary baking disaster by the distracted Anglo-Saxon monarch, and cramp balls, which describes the supposed value of pocketing the fruit bodies to relieve arthritis pain. *Xylaria polymorpha* is a related fungus, which is called dead man's fingers for the resemblance between its black fruit bodies and charred and swollen digits protruding from the ground.

Brown rot fungi decompose wood in a different fashion by breaking down cellulose without eliminating lignin. Lignin is modified to varying degrees by these fungi, but much of it remains after the cellulose is extracted and gives the powdery remnants of the wood a brown colour. Phylogenetic studies show that this decay mechanism evolved from the white rot form of decomposition. White rot fungi remove the lignin, but do not seem to derive any energy from its breakdown. They are simply clearing the polymer to access the cellulose. The development of the brown rot mechanism of decay, which bypasses lignin decomposition, may be a more efficient way of accessing the energy in cellulose. Brown rot basidiomycetes include *Laetiporus sulphureus*, sulphur shelf, which forms a bright yellow bracket. This is an edible mushroom that provides a splash of colour in stir-fry recipes. *Fomitopsis pinicola*, the red-belt conk, is another brown rot fungus that grows on dead conifers. The mycelium of this fungus can also colonize the wounded trunks and broken crowns of old trees causing stem decay. Stem decay produces cavities in large trees that become occupied by birds and mammals. The mechanism of wood decay by ascomycetes is different from rots caused by basidiomycetes and the effects of some of these fungi are described as soft rots rather than white or brown rots.

Most of the fungi that engage in wood decomposition are filamentous species rather than yeasts. *Daldinia concentrica* and other ascomycetes fashion a relatively restricted mycelium within decaying wood, but many of the basidiomycetes access dead trees from an expansive mycelium in the soil. The development of a large mycelium allows the fungus to transfer phosphorus and nitrogen from the soil and leaf litter to the sites of wood decomposition. This is vital because the levels of these nutrients in decaying wood are often too low to support growth of saprotrophic fungi. Transmission of water and nutrients across mycelia of some fungi occurs through channels within multicellular structures called cords and rhizomorphs. The term rhizomorph is reserved for the larger of these 'organs' that have a distinct rounded tip, like a plant root, that pushes through soil. Cords and rhizomorphs vary in complexity from bundles of hyphae that combine to produce thin cylinders, to fat pipes produced by the coordinated growth of hundreds of thousands of hyphae. Mycelia use cords and rhizomorphs to reallocate water and minerals between locations to optimize growth and decomposition.

Mycelia of different species of fungi or different strains of the same species compete for access to the cellulose in wood. When the combatants meet in the soil or within rotting timber they attack one another's hyphae by releasing toxins. These bouts can end with the destruction of one of the contestants, or in a deadlock in which hyphae of both mycelia stop growing in a zone of interference. In dying trees, these interactions are revealed as plates of dead hyphae that blacken with pigmentation. These plates can define territories occupied by a single fungus that runs lengthways along the trunk but is constrained from growing sideways by neighbouring fungi. A single tree can become filled with decay columns formed by interactions between different fungi. No scrap of wood is left alone. The patterns of dark lines in spalted wood are created by this process of mycelial competition. Every decorative bowl turned from spalted wood is a record of internecine warfare.

Cooperation between mycelia is another common phenomenon among individuals of the same species. This is manifested in the fusion of hyphae and the creation of large networks of hyphae and cords that can explore a wider area of the forest floor in search of food. Because fusion between hyphae in one location does not result in the immediate spread of nuclei of both mating types throughout the whole colony, large mycelia can grow as patchworks of genetic variation. Mycelia can also divide into two or more individuals when, for example, distant sections of a colony become detached from the older part of a mycelium and its mushrooms. All of these processes occur during the lives of the largest fungal colonies, including the enormous mycelia of *Armillaria solidipes* in Oregon described in Chapter 1.

Saprotrophic fungi utilize some of the same mechanisms of wood decay in forests to rot timber in buildings. Damage to timber frames in European homes is caused by the dry rot fungus, *Serpula lacrymans*, and *Coniophora puteana*, which causes wet rot. Another basidiomycete fungus, *Meruliporia incrassata*, is responsible for dry rot in the Western United States. The dry rot fungi grow from wet masonry or from soil outside buildings and can transport water and dissolved nutrients to the sites of wood decay through their cords. The cords of *Meruliporia* can be as fat as a garden hosepipe and grow for several metres. Once a mycelium of a dry rot fungus has colonized a wood beam, it forms long flattened fruit bodies that shed basidiospores into the air. These spores can initiate wood decay in other locations within a building. Wet rot is different because *Coniophora* shows a preference for growth on wood that is soaked with water. All three species cause the brown rot form of wood decay and are related to boletes whose mushrooms have pores rather than gills.

Indoor fungi, food spoilage, and decomposition of manufactured products

Many other saprotrophic fungi grow in buildings on shower curtains, plumbing fittings, and in drains and dishwashers. Diligent cleaning reduces the growth of fungi and other microorganisms in our homes, but they can never be abolished completely. A few of the fungi identified in buildings can cause serious illnesses, but it is important to recognize that their spores are common, if not ubiquitous, in the outdoor air. This means that the indoor environment does not pose any special danger. Ascomycetes are the commonest of the indoor fungi. When buildings are flooded, they grow on a wide range of surfaces including wallpaper and drywall, carpeting, and furniture. Under these circumstances fungal growth can cover large areas and produce high concentrations of spores in the indoor air. Spores carry proteins on their surface that can cause allergic responses that can be a serious problem for asthmatics. Allergies to fungi are featured in Chapter 7.

Saprotrophs also cause food spoilage in our homes, rotting fresh fruits and vegetables, and putrefying dairy and meat products. Species of the ascomycete *Penicillium* cause blue and green moulds of citrus fruits, blue mould of apples and pears, and spoil cheeses that are kept too long in a refrigerator. There is an irony with cheese spoilage because blue cheeses are flavoured by the growth of *Penicillium* species that differ from those associated with spoilage (Chapter 8). *Penicillium* is also responsible for the blue colonies that develop on mouldy bread. Black bread mould is caused by *Rhizopus stolonifer*, which is a zygomycete. The black dots that speckle the spoiled bread are sporangia filled with asexual spores. With a hand lens the sporangia are visible at the top of translucent stalks that project from the surface of the mycelium.

Other common ascomycetes that cause food spoilage are species of *Aspergillus* and *Alternaria*. *Aspergillus niger* contaminates baked goods, dairy products, and fruit juices. *Alternaria alternata* produces dark spots on tomatoes and other fruits and vegetables. Some of the fungi that cause food spoilage produce toxins, called mycotoxins, which can pose food safety concerns. Aflatoxins produced by *Aspergillus flavus* and *Aspergillus parasiticus* are the most significant mycotoxins from the perspective of human health. These cancer-causing compounds can reach potentially dangerous levels in cereals, peanuts, and other crops stored under hot, humid conditions. They can also be found in milk from cows that have eaten contaminated cattle feed.

More surprising than food spoilage is the growth of saprotrophic fungi on oil-based products. *Hormoconis resinae* is a filamentous ascomycete that can decompose hydrocarbons in aviation fuel. The fungus, accompanied with bacteria, grows at the interface between water droplets and the hydrocarbons in fuel. This microbial community damages rubber and plastic components in fuel systems and causes corrosion of the metal parts through the release of acidic compounds. Fuel contamination can also clog filters and damage fuel pumps. The problem was identified in the 1960s, when investigators found the fungus in a high percentage of several types of military and commercial aircraft. A solution came from the discovery that anti-icing fuel additives inhibited the growth of *Hormoconis*. These compounds are standard components of fuel used in aircraft today. *Hormoconis* is also known as the creosote fungus for its ability to grow on this toxic wood preservative. *Neolentinus lepideus* is another creosote-tolerant fungus. This species is a basidiomycete that forms mushrooms with scaly caps. It is called the train wrecker for its growth on old railway sleepers (ties), but there is no evidence that it has caused any accidents.

Compared with these acts of metabolic virtuosity, fungal damage to books seems rather facile. After all, most paper is made from wood pulp. Fungi grow on the page edges of closed books and are responsible for some of the foxing stains visible on open pages. Prints and paintings are also vulnerable to spoilage by fungi. These alarming phenomena are controlled by maintaining low humidity in libraries and museums. Tiny fungal colonies spot the emulsion on the surface of old photographic slides and their filamentous nature is apparent when they are magnified by projection. This anecdote may not make much sense to younger readers, but it provides another example of the relentless growth of fungi on man-made products.

Global significance of saprotrophic fungi

The global significance of saprotrophic fungi is apparent when we consider their overall contribution to the carbon cycle. Lignocellulose decomposition is the largest source of CO_2 emissions, exceeding the quantity of CO_2 released by burning fossil fuels by a factor of ten. This statistic does not diminish the importance of fossil fuel consumption in the modern disturbance of atmospheric chemistry. The release of CO_2 from decomposition has always been balanced by the uptake of the gas by plants and photosynthetic microbes. The consumption of ancient deposits of coal, oil, and gas since the industrial revolution, and at an accelerating pace, has provided a new source of CO_2 that is poorly accommodated by the natural carbon cycle. It is responsible for the 25 per cent increase in atmospheric CO_2 levels in the last fifty years.

The largest coal deposits were formed in the Carboniferous Period (360–290 million years ago), with the burial of immense quantities of the woody tissues of giant clubmosses, horsetails, and seed ferns that flourished in tropical wetland forests. An intriguing explanation for the fossilization of these plants is that the white rot fungi capable of destroying their remains did not evolve until the end of the Carboniferous. According to genetic investigations using a molecular clock calibrated with fossils (Chapter 1), there was a lag between the evolution of woody plants and the emergence of fungi that could decompose them. The first mycelia that secreted lignin-degrading enzymes enjoyed an abundance of food in the coal forests and their success is written in the decline in coal formation in the Permian Period. No other organisms have ever mastered the chemistry of lignin breakdown. We turn in Chapter 7 to fungi that feed on the less resilient tissues of the human body.

Chapter 7
Fungi in animal health and disease

Our microbiome

Yeasts are budding in the sebaceous oil on your scalp as you read this chapter. Fungi are growing on other parts of your skin too and in your mouth and nasal passages. They are thriving in the openings from your reproductive, urinary, and digestive systems, and they live in astonishing numbers inside your gut. These fungi that live on and inside us are part of the galaxy of microorganisms called the human microbiome. We interact with fungi in a more casual way too, by inhaling their spores from first breath to last gasp. Most of our associations with fungi cause no problems and some of them support our well-being. These harmless or supportive relationships can be upset, however, when the skin is damaged by a severe burn or if fungi are introduced into the body during surgery. Serious fungal diseases or 'mycoses' also develop when our immune defences are weakened by viruses, immune therapy after an organ transplantation, or by cancer treatments. The resulting infections can be difficult to treat. This chapter considers the fungi that populate the healthy human microbiome and the nature of fungal infections. We will also look at mushroom poisoning, and the effects of magic mushrooms on perception.

Skin and hair

The scalp offers food for fungi in the form of sebum, which is the fatty secretion from hair follicles, and keratin in hair and skin flakes. Metagenomic studies show that species of *Malassezia* are the dominant fungi on our heads, torso, and arms, and in our ears and noses. These yeasts are harmless residents and they probably help us by controlling the growth of other microorganisms. Our relationship with *Malassezia* is secured by the dependence of the fungus on fatty acid molecules in sebum. Other fungi manufacture their own fatty acids, but *Malassezia* lacks the enzyme that strings together small molecules into these storage fats. *Malassezia* causes dandruff and a more extreme inflammatory condition called seborrhoeic dermatitis. Both skin conditions can be treated with simple antifungal compounds containing zinc or selenium. *Malassezia* can also block skin follicles causing more widespread skin irritation and is responsible for pityriasis, displayed as a rash of discoloured skin patches.

Other kinds of skin disease are caused by filamentous fungi rather than yeasts. Ringworm is an umbrella term for superficial infections caused by dermatophytes. Dermatophytes are ascomycetes that invade skin, hair, and nails, breaking down keratin and triggering inflammation. The hyphae of dermatophytes grow in the outer layer of the skin where they feed on dead skin cells and rarely penetrate deeper layers. *Trichophyton* species are the commonest dermatophytes. In athlete's foot their colonies grow between the toes and can spread over the skin on the top and bottom of the foot in severe cases. On the scalp, hyphae invade the hair shaft and feed on the keratin filaments. When the fungus produces its spores, the hair shaft bursts and breaks off at the base leaving a black dot in the empty follicle. Ringworm occurs on other parts of the body and all of these infections are contagious, spreading from person to person via spores.

The cause of ringworm was not recognized until the 19th century when Robert Remak, a Polish investigator, examined infected skin crusts with a microscope and observed oval bodies (spores) and connecting threads (hyphae). He went on to use himself as an experimental animal by sticking an infected skin crust to his forearm. After two weeks, he developed an active skin infection that contained fungal cells. This act of self-sacrifice went a long way towards proving that ringworm was caused by a fungus. It is interesting that Remak's conclusions were published in the 1840s, predating Louis Pasteur's experiments on the Germ Theory of Disease by twenty years. But Remak was not the first scientist to demonstrate the causal link between microorganisms and disease. This honour goes to a French investigator, Bénédict Prévost, who published a series of radical experiments in 1807 showing that the smut fungus, *Tilletia caries*, causes 'bunt' or 'stinking smut' of wheat.

None of the dermatophyte infections are life threatening, and most respond to treatment with antifungal agents. Fluconazole is a popular medicine that is directed against an enzyme used to produce ergosterol, which is an essential component of the fungal cell membrane. It has relatively few side effects because animal membranes contain cholesterol rather than ergosterol. This specificity is the goal of antifungal drug discovery, because many of the medicines that are effective against fungi have damaging side effects. Terbinafine hydrochloride (Lamisil) is another medicine which is active against an earlier step in the pathway of ergosterol synthesis than fluconazole. It is very useful when it is applied to the skin to treat athlete's foot and other types of ringworm, but is associated with liver problems when it is taken orally for many months to treat stubborn toenail infections (onychomycosis).

Trichophyton species are very common skin residents even when there is no obvious ringworm infection, which means that they may operate as commensals, causing no benefit or harm to the host, as well as pathogens when they cause disease. *Malassezia* species, which are not considered to be dermatophytes, span the continuum of symbioses from mutualism (if the fungus curbs the growth of damaging microorganisms), to commensalism, and parasitism (when they cause dermatitis). Many of the microorganisms that flourish in our microbiome behave in this malleable fashion in their interactions with the human body. Fungi that live in the genitourinary tract and digestive system can cause more serious diseases than the members of the skin microbiome. The yeast called *Candida albicans* is the most important of these fungi.

Candida and opportunistic infections

Candida is a normal member of the gut microbiome and also grows in the mouth and in the female genitourinary tract. When these normal commensal relationships are disrupted, it causes superficial infections in the mouth and vagina, referred to as oral and vulvovaginal candidiasis. Oral candidiasis or 'thrush' is also common in babies and in adults with suppressed immune systems. More serious infections occur when the yeast spreads to internal organs causing systemic candidiasis.

The infection process begins when *Candida* adheres to the surface of host cells. Contact initiates a developmental switch in the fungus, which converts itself from budding yeasts into filamentous hyphae. The hyphal form is associated with the penetration of the host tissues. This occurs in two ways. The first mechanism involves some molecular trickery in which the fungus secretes molecules called invasins that cause host cells to take up the living pathogen. The alternative approach seems more direct, with the hyphae penetrating the underlying tissues by secreting enzymes to weaken proteins and applying turgor pressure as a source of physical force to push their way in (Chapter 2). As the infection develops, *Candida* can produce layered communities of yeasts and hyphae in tissues and on the surface of catheters. These are called biofilms. Biofilms can make it more difficult to treat an infection because the cells that are buried beneath the surface are protected from antifungal agents. In the severest infections, *Candida* spreads through the bloodstream causing fever, shock, kidney failure, and widespread clotting and bleeding (disseminated intravascular coagulation). Candidiasis caused by *Candida albicans* is the most common infection acquired in hospitals.

Proponents of the 'yeast syndrome' and '*Candida* complex' misrepresent scientific facts about this fungus, suggesting that it is responsible for all manner of complaints including anxiety, fatigue, headaches, weight gain, and depression. These often high-profile alternative health practitioners recommend dietary changes and various nutritional supplements to get rid of *Candida* and restore good health. People are misled by the real problems associated with superficial candidiasis into believing that the mere presence of the yeast is damaging. Candidiasis is a significant problem for many women, but suggestions for eradicating the fungus are misplaced and create a false sense of insecurity about our intimate relationships with microorganisms. Fungi grow on us, inside us, and are all around us. Whether we like it or not, fungi accompany us throughout our lives and help decompose our bodies when we die.

Candida and *Malassezia* are part of everyone's microbiome. Other fungi are detected on some people and not on others. In general, the immune system is exceedingly effective in limiting the growth of fungi beyond the skin. This is evident from the diversity of fungi that can invade our tissues when the immune defences are impaired. Rare systemic mycoses develop when vulnerable patients are colonized by fungi that live as saprotrophs in the environment. Zygomycete fungi that rot fruits and cause food spoilage in kitchens produce potentially lethal infections called mucormycoses in immunocompromised patients. Burn patients and people with uncontrolled diabetes are also vulnerable to these fungi. Colonies of filamentous hyphae invade soft tissues in the nasal passages and can grow within the walls of blood vessels and spread to the brain. Antifungal agents can be used to slow the progression of the disease, but mucormycoses are very difficult to treat. *Aspergillus fumigatus*, which thrives in the warmth of compost heaps, is another opportunist that starts as a lung infection (aspergillosis) and can spread throughout the body. Similar infections are caused by multiple species of fungi whose invasive hyphae are blackened with melanin pigment.

Other systemic infections

More common systemic mycoses are produced by a small number of opportunists whose life cycles alternate between distinctive growth phases in the soil and in animal tissues. The most important diseases caused by these fungi are cryptococcosis, histoplasmosis (Ohio valley disease), blastomycosis, and coccidioidomycosis (valley fever). Cryptococcosis is caused by *Cryptococcus neoformans* and *Cryptococcus gattii*. These are species of basidiomycete yeasts that infect the nervous system. Outside the body *Cryptococcus* grows in the form of filamentous colonies in bird droppings. Human infections begin when spores or yeast cells produced by this saprotrophic phase are inhaled deep into the lung. In the mucous lining of the lung the fungus is consumed by macrophages and other cells of the immune system programmed to destroy microorganisms. But the fungus subverts this defence mechanism by sidestepping digestion inside the macrophages and using the host cells as vehicles for spreading around the body and into the bloodstream. Transmission from the bloodstream into the brain may involve uptake by another population of macrophages that have been aptly described as Trojan horses.

Once inside the brain tissues, *Cryptococcus* forms abscesses and produces a severe form of meningitis. Amphotericin B is the antifungal drug used against the most serious cases of cryptococcosis. Rather than inhibiting ergosterol synthesis like fluconazole, amphotericin B works by binding to ergosterol after it has been incorporated into the fungal membrane. This damages the membrane, making the fungal cells leaky. Unfortunately, amphotericin B also attaches to cholesterol, weakening the membranes of human cells and causing kidney damage. The way that *Cryptococcus* avoids destruction by macrophages may have evolved from ancient interactions with soil microorganisms that enabled the fungus to avoid digestion by amoebae. Based on this theory, the pathogenic behaviour of the fungus may be rooted in ecological relations that evolved long before the origin of mammals.

Histoplasma capsulatum is another fungus with distinctive saprotrophic and pathogenic phases. Like *Cryptococcus*, *Histoplasma* grows in soil enriched with bird droppings. It is also found in soil on the floors of caves that are fertilized with bat guano. The fungus grows in the lungs and only causes any serious symptoms when the immune system is compromised. AIDS patients, for example, are susceptible to rare cases of systemic disease when *Histoplasma* spreads to the spleen, liver, and adrenal glands. The common name of Ohio Valley disease comes from the concentration of cases along the Ohio and Mississippi river valleys. Blastomycosis is a similar illness caused by *Blastomyces dermatitidis*. It is distributed across a wide region that includes the Eastern United States and Canadian provinces as far west as Manitoba. Another species, *Coccidioides immitis*, is a soil fungus in the southwestern United States and northern Mexico that grows in the lung and can spread to other tissues and produce brain inflammation. Outbreaks of the disease have been linked to soil disturbance caused by farming, construction projects, and earthquakes, which release *Coccidioides* spores into the air. The common name for this mycosis is valley fever.

Most serious cases of fungal infection occur in people with compromised immune systems, but disease outbreaks do occur occasionally in people with apparently healthy immune systems. Immune responses to fungi range from mechanisms of general exclusion to attacks upon specific pathogens engaged in an ongoing infection. The skin and the mucous-covered surfaces of the lungs, gut, and genitourinary tract provide the first obstacles to fungal invasion. In addition to acting as physical barriers, antimicrobial compounds secreted by these tissues and resident commensal microorganisms interfere with pathogen growth. If these obstacles are breached, a series of innate or non-specific immune mechanisms destroy the foreign cells.

Chemical signatures that are common to all fungi alert the innate immune system. These include molecules that form the walls of yeast cells and filamentous hyphae. Recognition triggers an inflammatory response in which neutrophils, macrophages, and other kinds of white blood cells attack and destroy the pathogens. The innate response also involves the uptake and disassembly of yeasts, hyphae, and spores by dendritic cells. Dendritic cells process molecules released during the disintegration of the fungi and display these antigens on their surfaces. This action by the dendritic cells triggers the adaptive immune system that provides a specific response to a particular fungus. Antigen presentation activates T cells that coordinate the destruction of the fungal cells in other parts of the body. Populations of T cells called CD4+ helper T cells are vital to the operation of the adaptive immune system. Low levels of these T cells in AIDS patients and people coping with other forms of immune damage are an indicator of their susceptibility to opportunistic infections.

A form of pneumonia caused by a yeast called *Pneumocystis jiroveci* illustrates the importance of immunodeficiency in fungal infection. *Pneumocystis* pneumonia, or PCP, was an exceedingly rare illness before the beginning of the AIDS epidemic in the 1980s. Cases of this infection were so unusual that requests by physicians in California and New York for the antibiotic called pentamidine used to treat PCP were one of the first indications of the new epidemic. Reports of Kaposi's sarcoma, a rare form of cancer caused by a herpes virus, were another sign of an emerging disease that crippled the immune system. Eighty per cent of AIDS patients developed PCP before effective HIV medications were developed and it remains a major cause of death among AIDS patients today.

Our understanding of the biology of the fungi that cause opportunistic infections has advanced in ways that were not imagined a few decades ago. Molecular genetic experiments, the sequencing of whole genomes, and advances in microscopy have revealed masses of information about the operation of these microorganisms. On a clinical level, however, the modern exploration of pathogenic fungi has been disappointing. The quest for the molecular determinants of virulence has shown how *Candida* sticks to host cells and how *Cryptococcus* evades the immune defences, but these discoveries have not led to new treatments. The pace of antifungal drug discovery has been sluggish and the emergence of *Candida* strains that are resistant to fluconazole is a troubling development. The most serious mycoses continue to be treated with amphotericin B that was introduced in the 1950s. The most definitive approaches to treating these illnesses are likely to come from therapies that resolve the underlying immunological deficiencies that invite opportunistic infection.

Infections of other animals

Fungi that infect humans produce similar diseases in our mammalian relatives. Ringworm is common among domesticated dogs and cats, candidiasis and cryptococcosis occur in pets and farm animals, and coccidioidomycosis is encountered in a wide range of zoo animals. Aspergillosis is the most common respiratory infection in birds and a few of the human mycoses also occur in reptiles. Populations of wild animals are also prey to these fungi. Marine biologists have reported outbreaks of cryptococcosis and coccidioidomycosis among porpoises and dolphins, sea lions, and sea otters. The spread of highly virulent mycoses among single animal groups has become a major concern in our time. These include chytridiomycosis, caused by *Batrachochytrium dendrobatidis*, which affects one-third of amphibian species, white-nose disease (*Pseudogymnoascus destructans*) that has killed six million bats in North America, and marine aspergillosis of sea fan corals in the Caribbean caused by *Aspergillus sydowii*. Degradation of the environment by human activities may be a factor in some of these epidemics, with climate change, ozone depletion, and multiple pollutants implicated in various studies. Each of these diseases is so complex, however, that it has been impossible to identify a single cause.

The rapid spread of unusual fungal diseases is a new and disturbing phenomenon, but mycoses are a normal part of animal life and death. Just as all animals engage in mutualistic and commensal symbioses with fungi (Chapter 4), pathogenic fungi affect every species. Pathogenic interactions are very evident among the insects. Insects and spiders are attacked by *Cordyceps* species that penetrate their exoskeletons, invade their soft tissues, and break out from the dead animal with club-shaped fruit bodies. *Cordyceps militaris* infects the larvae and pupae of moths and butterflies, and transforms the host into masses of white hyphae. This process is referred to as mummification. If the insect dies in soil or rotting wood, the fungus is recognized from its orange clubs protruding into the air. The largest of the fruit bodies are the size of pencil stubs. The infectious ascospores of *Cordyceps militaris* are long filaments. These are ejected into the air flowing around the top of the club and fragment into multiple infectious particles.

Related species that infect adult insects modify the behaviour of their prey, stimulating them to climb to the top of grass stalks or to crawl on to leaves. As the infection proceeds, the fungus sprouts from the dead insect and its fruit bodies are in a perfect position to ensure wind dispersal. The biochemical mechanism used by the fungus to control the behaviour of the host has not been solved. A particularly fascinating species of *Ophiocordyceps* that parasitizes ants has been studied in tropical rainforest in Thailand. After infection, the ants develop convulsions and adopt a 'drunkards walk'. These symptoms are observed in ants that have fallen from their preferred habitat high in the tree canopy. In the humid atmosphere of the forest understorey, these 'zombie ants' crawl on to saplings and bite into the main vein on the underside of leaves. This is synchronized to the elevation of the sun, or to a related environmental variable like temperature or humidity, because the infected ants always bite leaf veins around noon. Meanwhile, the fungus destroys the muscles controlling the mandibles, which locks the insect in its 'death grip'. The insects live for up to six hours after they bite into a leaf, twitching their legs but unable to release themselves. After a few days, the fruit body of the fungus erupts from the head of the ant and discharges its filamentous spores (Figure 26). Characteristic bite marks have been found on 48 million-year-old fossilized leaves showing that this sophisticated relationship between fungi and ants is a very ancient form of parasitism.

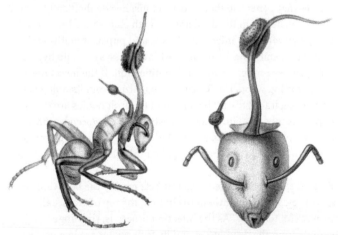

26. *Ophiocordyceps unilateralis* on rainforest ant.

Allergens, hallucinogens, and poisons

Fungi have adverse effects on animal health besides their role as pathogens. Asthma is a serious global health problem that affects 300 million people and is implicated in 250,000 deaths each year. Fungi are a significant cause of asthma. The frequency of asthma symptoms fluctuates in response to seasonal changes in the concentration of airborne spores, and hurricanes and other climatic events worsen the problem through their widespread effects on spore dispersal. Allergy to fungi results from interactions between the immune system and proteins carried on the surface of the commonest kinds of spores. Dendritic cells in the lungs process spore fragments in the same way that they handle pathogens, but the cascade of immunological reactions in people with allergies results in the release of histamine and other inflammatory compounds that produce the symptoms of asthma and allergic rhinitis (called hay fever when it is seasonal).

There has been a lot of anxiety, particularly in the United States, about the purported toxicity of fungi that grow in flooded homes. The spores of some of these indoor fungi, including a black-pigmented ascomycete called *Stachybotrys chartarum*, carry toxins that can cause a range of symptoms if they are absorbed in high concentrations. Most of the available evidence suggests that people that inhale spores of this fungus in water-damaged buildings are not exposed to levels of these mycotoxins that can cause illness. Nevertheless, the inhalation of large quantities of allergenic spores in these circumstances is a serious public health issue. Fungal growth inside buildings is indicated if the concentration of spores in indoor air exceeds the concentration measured in outdoor air on the same day. The detection of different kinds of fungi in indoor and outdoor air is another sign of the active growth of fungi inside buildings. *Stachybotrys* and the other microscopic fungi that can thrive in flooded buildings are referred to as indoor 'moulds', but this term, like 'mildew' and 'toadstool', has no precise scientific meaning.

Ergotism, resulting from the consumption of rye bread contaminated with ergot, was a major threat to public health in the Middle Ages. The ergot fungus, *Claviceps purpurea* (Chapter 4), is an ascomycete pathogen that infects the flowers of cereals and replaces the seeds with hard black nuggets called sclerotia. Rye flour milled from infected crops was responsible for documented outbreaks of ergotism in France and other European countries between the 6th and 12th centuries. Symptoms of poisoning included severe burning sensations in the limbs due to the constriction of blood vessels, convulsions, and hallucinations. Gangrene resulting from vasoconstriction was often lethal and the affliction became known as St Anthony's Fire. Ergotamine is one of the poisonous alkaloids in ergot. Ergotamine in combination with caffeine is prescribed as a treatment for migraine headaches today. Ergotamine is also used to synthesize lysergic acid, which is a precursor of the drug LSD or 'acid'. The hallucinogenic effects of LSD were discovered in 1943 by Albert Hofmann who was investigating the pharmacological properties of ergot extracts. Another ergot alkaloid, ergometrine, is used to ease delivery of the placenta and to prevent bleeding after childbirth.

Widespread recreational use of LSD in North America and Europe began in the 1960s, at the same time that psychedelic or 'magic' mushrooms became a symbol of anti-establishment counterculture. The most potent psychoactive compound in these fungi is psilocybin, which is produced by more than 200 mushrooms including the *Psilocybe* species. Psilocybin is converted to psilocin after the mushrooms are eaten. The chemical structure of psilocin is very similar to serotonin, which is a neurotransmitter associated with feelings of happiness. When psilocin binds to serotonin receptors in the brain, the disturbance to neurological function causes euphoria and a range of hallucinatory experiences.

Clinical studies on test subjects given purified psilocybin reveal that the drug causes a temporary reduction in blood flow to certain parts of the brain that diminishes neurological activity.

Stimulation of crosstalk between regions of the brain that do not normally communicate is another effect of psilocybin (psilocin). This 'synaesthesia' can result in the perception of colours when listening to music. Other common experiences include visions of geometric patterns, altered perception of the passage of time, and stimulation of memory. Some people who consume magic mushrooms have very frightening experiences, but the frequent positive responses to psilocybin have encouraged studies on its use to treat depression. Different compounds in fly agaric mushrooms (*Amanita muscaria*) raise serotonin and dopamine levels and induce a range of hallucinatory experiences.

Mushroom poisoning is a terrifying consequence of picking the wrong species when people are searching for magic mushrooms or for edible species. The horror of these life-changing and life-ending errors is one of the reasons that mycologists with skills in mushroom identification are so valuable. The most potent mushroom toxins include amatoxins, produced by the death cap (*Amanita phalloides*), which cause liver failure, and orellanine and cortinarins, found in webcaps (*Cortinarius* species), which attack the kidneys and liver. Gyromitrin is a toxin produced by a species of false morel, *Gyromitra esculenta*. False morels are ascomycetes. Their fruit bodies have stalks and convoluted heads that resemble the shapes of 'standard' basidiomycete mushrooms, but their spores are dispersed from the exposed surface of the cap rather than falling from protected gills. When these mushrooms are eaten, gyromitrin is converted into monomethylhydrazine which is a chemical used in rocket fuel. Nausea and vomiting are the usual symptoms of poisoning, but organ damage has been reported in a few cases and some victims have died. Despite the risk, false morel aficionados enjoy eating this fungus after boiling the fruit bodies and discarding the water to remove the toxin.

The biological reasons that fungi produce these remarkable hallucinogens and toxins are not understood. It seems likely that some of these molecules work as pesticides that protect fruit bodies against insects and other invertebrates. Humans, in other words, are not the intended target for these compounds. Very few of the 16,000 mushroom species are poisonous and a few, as most readers will agree, combine safety with delicious flavours. In Chapter 8 we turn to the pleasures of edible mushrooms and other ways in which fungi are used to improve our lives.

Chapter 8
Edible mushrooms and fungal biotechnology

It is appropriate to end this short book with a chapter on the uses that we have made of our mycological wisdom. The pleasure of eating wild mushrooms, tempered with awareness of the poisonous nature of a few species, is an ancient experience born from our history as hunter-gatherers in forest ecosystems. Cultivation of edible mushrooms is among the oldest biotechnological uses of fungi, although brewing and baking with yeast go back even further. Until the 19th century, beer, wine, and bread were made by people who utilized the physiological activities of fungi without knowing that microorganisms existed. Artisanal uses of fungi in cheesemaking and the fermentation of a fabulous array of Asian foods are similar instances of the practical uses of fungi in the absence of scientific knowledge. From these long-standing interactions between humans and fungi, we come to the modern intentional application of fungi for the production of antibiotics and a plethora of other medicines, fermentation of meat substitutes, synthesis of industrial enzymes and acids, and manufacture of biofuels.

Picking wild mushrooms

People pick wild mushrooms in the pursuit of mycological knowledge, to jazz up their diet, and to make money.

Mushroom identification is a tricky business and there is no substitute for learning from a mycologist with years of practical experience. Some species are more difficult to identify than others. The magnification of specimens with a hand lens is often helpful and indoor study ranges from the collection of spore prints from mushroom caps and detailed examination of mushroom tissues and spores using a microscope. Photographic records are also important. Academic study involves molecular identification and the collection and drying of specimens for deposition in a herbarium. Mycological societies play an important role in teaching people to identify mushrooms and this often involves gathering large numbers of fruit bodies and their display on tables. This seems wasteful when the haul of shrivelling fungi is thrown away at the end of the day, but this is more of an aesthetic concern than an issue of sustainability. Mushroom picking for cooking is more popular in some cultures than others and there is an overlap between the study and consumption of wild mushrooms among amateur mycologists.

Trade in wild mushrooms occurs in many regions, with markets in France and elsewhere in Europe for a range of edible species including ceps or porcini (*Boletus edulis*), girolles (chanterelles, *Cantharellus cibarius*), and the beautiful orange Caesar's mushroom (*Amanita caesarea*). Specialized seasonal markets sell white truffles (*Tuber magnatum*) in northern Italy and black Périgord truffles (*Tuber melanosporum*) in southern France. Restaurateurs bid for these delicacies in international auctions. Species of *Terfezia* and *Tirmania* are desert truffles that grow in association with *Helianthemum* bushes in semi-arid and arid deserts from Morocco to Saudi Arabia. These expensive 'mushrooms' are very popular in North Africa and the Middle East. The reproductive mechanisms of fungi that rely on the production of vast numbers of spores by short-lived fruit bodies encourages the idea that mushroom picking is a harmless pastime. There are some limits to this assumption.

When mushroom picking becomes a commercial activity that encourages the wholesale removal of fruit bodies from particular areas every year, the sustainability of the practice is questionable. A thirty-year experiment on the impact of mushroom picking in Switzerland showed that intensive picking had no effect on subsequent crops as long as the pickers worked from a catwalk. This indicated that mycelia can be very resilient. But because mushroom spores are dispersed by wind, the loss of the genetic diversity derived from fruit bodies in one patch might have wider consequences that are not seen in a study confined to one area. There are more immediate concerns raised by the Swiss study. When the investigators left the catwalk to pick normally, fewer mushrooms were formed in subsequent years in the trampled areas. This means that intensive picking has the potential to do serious damage to mushroom populations.

Mushroom harvesting in the Pacific Northwest of the United States is one area of concern, and in China, where annual exports of wild mushrooms may run into the hundreds of thousands of tons of fruit bodies, the future of the 'industry' seems bleak. *Ophiocordyceps sinensis* is an interesting case. This parasite of moth caterpillars is related to the zombie ant fungus described in Chapter 7. It fruits in alpine meadows in the Tibetan Plateau and Himalayas and is used in Chinese and Tibetan medicine. Trade in the 'caterpillar fungus' supports a global market valued between US$5 billion and $11 billion. One study showed that the harvest in Nepal fell by 50 per cent between 2009 and 2011. Management of this natural resource is complicated by the reliance of many people on the fungus as a source of income. The same issue is raised in the Pacific Northwest as a defence of people who gather chanterelles, boletes, and matsutake mushrooms. Without management, however, these mushroom grounds, like many commercial fisheries, will collapse.

Mushroom cultivation

Mushroom cultivation offers a sustainable source of fruit bodies of a limited number of domesticated fungi. The white button mushroom, *Agaricus bisporus*, is the most popular species, followed by shiitake, *Lentinula edodes*. *Agaricus bisporus* is grown on straw blended with animal manure and gypsum. This mixture is composted by bacteria and then pasteurized by steaming before inoculation with mushroom spawn. Mushroom spawn is prepared by growing *Agaricus bisporus* on cereal grains. On modern farms, beds of compost are supported on metal shelves in growing rooms in which the environmental conditions are controlled to optimize mycelial development. After two to three weeks, fruiting is stimulated by casing the compost with a layer of peat and limestone, lowering the temperature in the growing rooms, and increasing airflow to drop carbon dioxide levels. Mushrooms appear in a few weeks and three flushes can be picked from a typical bed.

Agaricus bisporus is the mycological equivalent of the banana, an agricultural oddity with too little genetic diversity. Unlike wild species, mycelia that grow from single spores of the white button mushroom can form a new generation of fruit bodies without mating. Reproduction is clonal, which is a good thing from the perspective of product consistency but worrying from the viewpoint of pest resistance. Mushroom breeders have had some successes in strain development and different varieties of this fungus have been produced that form darker crimini mushrooms rather than white buttons. Older fruit bodies of the crimini strains are sold as portabellos.

Shiitake mushrooms are grown on slender hardwood logs arranged in stacks or propped against one another. Plugs of shiitake spawn are pressed into holes drilled into the logs and sealed with wax. If everything goes well, the mycelium will rot the wood over the following six to eighteen months and crops of mushrooms can be harvested for up to five years from a productive stack of logs. Shiitake was grown in this fashion in China for more than 1,000 years before it became popular in Western countries. China 'owns' the cultivated mushroom market today, producing 90 per cent of the world's shiitake and more than one-third of white button mushrooms. Other cultivated mushrooms that account for slivers of the market share include the wood ear (*Auricularia auricula*) and silver ear (*Tremella fuciformis*). Both of these species are wood-rotting basidiomycetes that form jelly-like fruit bodies. They are sold in a dried form and are popular in Asian cuisine.

Medicinal mushrooms and antibiotics

Mushrooms have been used as medicines since the Neolithic. Two species of bracket fungi, or polypores, threaded on strips of animal hide, were found on the mummified body of 'Ötzi the iceman', who died 5,300 years ago in the Alps. One of these fruit bodies was probably used as tinder, but the dried chunk of the birch polypore, *Piptoporus betulinus*, may have been carried for its medicinal properties. Today's controversy surrounding the use of medicinal mushrooms arises from conflict between the traditions of East Asian medicine and the evidence-based medicine of the West. Shiitake, for example, has been used since the Ming Dynasty as an energy boosting 'tonic' that is recommended for the elderly and for patients convalescing after serious illnesses. It is also prescribed as a specific treatment for respiratory disease, liver disease, and parasitic worms.

The absence of critical experimental support for the effectiveness of shiitake in any of these applications is of little interest to advocates of 'Chinese' medicine. They argue that their approaches to patient care are so different from those practised in modern medicine that they cannot be assessed using the same standards. There is some truth to this objection. Prescription drugs used in

Western medicine contain one or a few active ingredients and are subject to testing in placebo-controlled clinical trials. It is difficult to apply the same criteria for medicinal mushrooms because they may contain many active ingredients whose levels differ from sample to sample. Yet without clinical trials, physicians and patients have no basis for using mushrooms other than trust in the stories told by people who claim to have benefited from their use. And one of the problems with this faith in anecdote is that companies that market mushroom extracts prey upon consumers by making unsubstantiated claims. This is particularly distressing when people with serious illnesses are fooled into believing that mushrooms can compensate for the inevitable limitations of modern medicine.

Putting aside the claims about miracle cures offered by mushroom extracts, there are good reasons for believing that fruit bodies do contain compounds with powerful pharmacological properties. The toxins of death caps and web caps are obvious examples of molecules that have a powerful effect on human physiology and mushrooms produce a treasure trove of metabolites of unknown function that might have beneficial medical applications. Most of the research on the properties of specific mushroom extracts has been performed on cultured cells and on laboratory mice. These studies have shown that cell wall components from shiitake and the turkey tail mushroom (*Trametes versicolor*) stimulate cells in the immune system and may have some protective effects against tumour growth. Experimental treatment of cancer patients with these compounds has had mixed results, but these limited trials certainly encourage further investigation. The combined use of mushroom extracts with established chemotherapeutic agents has produced some of the most encouraging results.

While the medicinal value of mushrooms is unproven, other fungi are a verified source of 'miracle' drugs. The discovery and development of antibiotics synthesized by filamentous fungi are triumphs of Western medicine. In a time of increasing concerns about antibiotic-resistant bacteria it is easy to forget how the discovery of penicillin in 1928 by Alexander Fleming changed our lives. Fleming shared the Nobel Prize in Physiology or Medicine in 1945 with Howard Florey and Ernst Chain, who developed penicillin for clinical use. Florey and Chain led a research group at Oxford University and tested the antibiotic on the first patient in 1941. Penicillins are secreted by species of the ascomycete *Penicillium* and belong to a class of antibiotics that share a common motif called the β-lactam ring. These compounds work against bacteria by weakening their cell walls, which causes them to burst. Resistant bacteria produce enzymes that inactivate the antibiotics by opening their ring structure.

Penicillin is produced by *Penicillium chrysogenum* grown on lactose (carbon source) and yeast extract (nitrogen source) in industrial fermenters. Synthesis of penicillin rises as the fungus depletes the lactose and production is stimulated by aeration and agitation. Meticillin and ampicillin are examples of semisynthetic antibiotics produced by the addition of side chains to the core ring structure of penicillin. Meticillin (formerly methicillin) has been replaced with other antibiotics with the emergence of meticillin-resistant strains of bacteria including *Staphylococcus aureus* (MRSA). Ampicillin remains effective against a wide range of bacteria. Cephalosporins are β-lactam antibiotics produced by another ascomycete, *Acremonium chrysogenum*. These are used as the last line of defence against MRSA. Filamentous fungi are a source of many other pharmacological agents. Lovastatin is a cholesterol-lowering drug that is produced by *Aspergillus terreus*. This is marketed as Mevacor® and a synthetic derivative is trademarked as Zocor®. Cyclosporin A is synthesized by a soil fungus called *Tolypocladium inflatum*. This compound suppresses the immune system and is used to support patients following bone marrow and organ transplantation. Cyclosporin is also an effective treatment for psoriasis, severe dermatitis, and rheumatoid arthritis.

In addition to their importance in pharmaceutical synthesis, fungi are a key source of industrial enzymes that are used in food and beverage industries, personal care products, laundry detergents, textiles, and leather manufacture. Fungal cellulases and phytases are used in the production of livestock feed, and laccases are added to breath fresheners to destroy the compounds responsible for halitosis. Laundry detergent formulations contain fungal cellulases that increase the softness and brightness of clothing. These enzymes work in cold water, which saves energy, and their effectiveness at removing dirt reduces water consumption. Other products containing fungal enzymes are effective at bleaching raw cotton before it is dyed, whitening paper, and removing gums from natural fibres.

Genetic engineering has become very important in fungal technology. Various methods are used to transform fungi with foreign genes so that they produce enzymes that are not part of their natural repertoire of proteins. This is necessary because some species that are sources of valuable proteins do not grow well in fermentation vessels or do not produce commercial levels of their products. If another species can be coerced to make more of the foreign protein it may serve as a very profitable living factory. *Trichodermi reesei* is a cellulose-decomposing fungus that has been transformed in this fashion to generate high yields of enzymes that are used in many industries. Organic acids are another fungal product category with wide applications. *Aspergillus niger* is used to produce citric acid, which is used as a preservative and food flavouring, and is an additive to water softeners and fertilizers. Gluconic acid and itaconic acid produced by *Aspergillus* and *Penicillium* species have applications in metal finishing, cement manufacture, and polymer synthesis.

Brewing and baking

The industrial applications of fungi are recent innovations compared with the ancient uses of yeast in brewing and baking. Brewing and baking are the original biotechnologies.

Humans have brewed beer from cereal grains and made wine from fruits and tree sap for millennia. Some archaeologists have proposed that cereal cultivation was driven by our species' love of beer rather than any pressing need to grow grains for their nutritional value. Whether beer came before bread, or vice versa, the associated development of cereal cultivation encouraged human settlement. The earliest brewers relied on fermentation by wild yeasts that bloomed on the grains and fruits used in their recipes. Domestication of *Saccharomyces cerevisiae* began when brewers developed methods for deliberate transfer of the fungus from one fermentation to the next. With the introduction of this practice, yeast strains were set on novel evolutionary tracks where they were provided with an abundance of food in return for their effectiveness at producing alcohol. Genetic profiling of yeast strains suggests that this began more than 10,000 years ago. Contrary to the general pattern of genetic isolation of agricultural plants and animals, resulting in very limited genetic variation, yeast strains used in beer brewing and winemaking continue to vary according to geography.

Saccharomyces cerevisiae and other yeasts obtain their energy from sugar metabolism and generate carbon dioxide and water as waste compounds when oxygen is available. This is aerobic respiration and provides the greatest energy yield from the consumption of sugars. When oxygen levels fall, or if the yeast is growing in very high concentrations of sugar, cellular metabolism is redirected to a different pathway that releases carbon dioxide and ethanol. This fermentative mechanism is less efficient than aerobic respiration and leaves a lot of 'unburned' calories in ethanol. This simple switch by the yeast to an inefficient burn is the foundation of the great variety of alcoholic drinks.

The basics of brewing and winemaking are covered in so many books and online sources that it seems more interesting to consider the role of fungi in other brewing practices in this chapter. Apple cider and palm wines have huge regional significance. Cider apples are classified according to sweetness, ranging from bittersharp (high in acidity and tannin) to sweet (low acidity and low tannin). The apples are crushed and the pulp is pressed to release the juice. Modern 'industrial' cidermaking involves inoculation with specific yeast strains and addition of sulphur dioxide to exclude other microorganisms. Traditional cidermaking relies upon yeasts carried on the apple skin or clinging to the machinery between pressings.

Studies of traditional cider factories show that numerous yeast species are involved in the fermentation. Yeasts associated with the harvested apples multiply in the juice for the first few days and drive the fermentation until they are overcome by the rising levels of alcohol. As the number of these pioneers declines, the community of fungi changes and *Saccharomyces cerevisiae* becomes the principal species. This is explained by the higher alcohol tolerance of *Saccharomyces*. A third blend of yeast species controls the final maturation phase of cidermaking. Similar successions of microbial communities occur in traditional fermentations that produce other alcoholic drinks. Bacteria accompany yeasts in many fermentations. Lactic acid bacteria convert the tart-tasting malic acid into the softer lactic acid in cider, beer, and wine, but can also spoil fermentations by producing unpleasant flavours.

Palm wines produced by fermenting the sap from palm trees are common in many countries in Africa and Asia. These potent brews are fermented by complex communities of yeasts and bacteria, but wild strains of *Saccharomyces cerevisiae* always seem to dominate these drinks. Fifteen different strains of this ubiquitous yeast were identified in a metagenomic study of palm wine from Cameroon. Palm wines are among the oldest alcoholic drinks, whose discovery was an inevitable consequence of the attraction of natural populations of wild yeast cells to palm sap. Yeasts operated in the same way long before human evolution. We find, for example, that mammals that pollinate palm trees drink natural alcoholic nectar fermented in the inflorescences of these plants. Yeast has been making alcohol for animals for millions of years.

Saccharomyces cerevisiae has been used to ferment dough made from wheat and rye flour for more than a millennium. Froth from beer vats filled with top yeast was used for breadmaking by Romans and this method was widespread in the 19th century. By starting fresh dough with a small quantity of the mix saved from the previous batch of bread, called leaven in the Bible, bakers engaged in the unconscious selection of vigorous yeast strains. The first commercial food yeast was produced in Holland and sold as a cream in the 18th century. This was superseded by compressed cakes of yeast in the 19th century and granulated dry yeast became the choice of bakers in the 1940s. Baker's yeast is manufactured today using fed-batch fermentation with molasses supplemented with a variety of nutrients. Fed-batch fermentation allows the producer to add nutrients to the reaction intermittently, or continuously, to control the metabolic activity of the yeast and generate high cell densities. These methods are used to produce 200,000-litre batches of starter yeast in huge stainless steel fermenters that fill entire buildings. The ethos is quite different for the growing number of artisanal bakers who exchange small quantities of starter dough with their friends.

Cheesemaking involves many microorganisms, beginning with bacteria that ferment lactose in milk. Yeasts and filamentous fungi play subsidiary roles in flavouring and ripening cheeses. The white rinds on Brie and Camembert are formed from the dense white mycelium of *Penicillium camemberti*. *Penicillium roqueforti* proliferates in blue-vein cheeses including Roquefort, Gorgonzola, Stilton, and Danish blue. Fungi play a primary role in the fermentation of traditional Asian foods. Tempe, originating in Java, is manufactured by inoculating cooked soybeans with spores of the zygomycetes *Rhizopus* and *Mucor*. A similar method is used to produce furu or sufu, which is a cheese-like food made from soybeans in China. Soy sauce is made by inoculating a mixture of boiled soybeans and roasted wheat with *Aspergillus* species to form 'koji'. After a few days, the koji mash is mixed with brine to produce 'moromi', and this is fermented by yeasts and bacteria for several months to produce this staple condiment of Asian cuisine.

Quorn is a popular meat substitute produced by *Fusarium venenatum* that is grown in connected pairs of fifty-metre tall air-lift fermenters that hold 230 tons of broth. The vessels are connected at the top and the bottom to form a continuous loop. Compressed air and ammonia are pumped into the bottom of the first vessel, called the 'riser', oxygenating the culture and circulating the liquid containing the fungus towards the top. Carbon dioxide from the respiring fungal cells is released through a vent, and the liquid falls through the second 'downcomer' vessel, where it is infused with fresh nutrients. A heat exchanger at the bottom of the system maintains the temperature at 30°C and the culture is harvested at a rate of thirty tons per hour. Quorn production is completed by heating, drying, mixing with egg white, and the addition of flavourings and colourings.

Bioethanol and bioremediation

Saccharomyces cerevisiae is used to generate bioethanol from sugar cane in Brazil and from corn in the United States. Juice separated from sugar cane fibre is concentrated to make sugar and molasses. The fibre is burned as a source of energy for the biofuel plant, the sugar is refined for the food industry, and ethanol is produced from the molasses. Corn is a more complicated 'feedstock' for bioethanol production because its seeds are rich in starch rather than sucrose and other sugars. This adds a step to the production process, because the starch must be converted into sugars using enzymes. Some of these enzymes are fungal products. The efficiency of bioethanol production would be revolutionized if fungi could be used to release sugars from agricultural waste containing fibrous lignocellulose polymers. Natural decomposition of woody plant debris by white rot basidiomycetes is the model for the development of this industrial process. Pilot studies in which rice straw is milled and fed to mycelia of the oyster mushroom, *Pleurotus ostreatus*, and other white rot fungi are promising, with significant breakdown of lignin, cellulose, and other polymers. The next step in this 'second generation' biofuel production is the addition of yeast to ferment the sugars into ethanol. Agricultural wastes would provide a limitless carbon-neutral supply of fuels, but we are decades away from enjoying the benefits of this emerging technology.

White rot fungi also have significant potential as decontamination agents in soils polluted by oil and gas extraction, the mining industry, chemical companies, and agriculture. Interest in this field, called 'bioremediation', is encouraged by the effectiveness of the wood-rotting enzymes secreted by fungi at breaking down toxic organic compounds. Decomposer basidiomycetes are able to decontaminate wood chips impregnated with hydrocarbons and chlorophenols in experimental trials. Other filamentous fungi and yeasts degrade pesticides, dyes, toxic solvents, and explosives in culture. Scaling up these processes for larger clean-up projects is a considerable challenge for investigators.

Appreciation of the importance of fungi can lead to wishful thinking about the ability of these extraordinary microorganisms to counteract the environmental damage caused by our species. Because fungi sustain forest trees through mycorrhizal interactions and by enriching the soil through wood decomposition, it is tempting to think that they can restore productive ecosystems after the original habitats have been ruined. This is a simplistic view of ecology. Human civilization is supported by the biological activities of the fungi that have been introduced in this book, but there are evident limits to the ability of the fungi to 'save the planet'. In closing this short introduction to mycology, I leave you with two unassailable facts: fungi are everywhere, and will outlive us by an eternity.

Further reading

Academic journals

Mycology is such a vibrant area of research that any book that aims to be current is certain to miss important findings within a few months of publication. Academic journals are the best source of current information and the following periodicals showcase current mycological research: *Fungal Biology*, *Fungal Biology Reviews*, and *Fungal Ecology* are published by the British Mycological Society (<http://www.britmycolsoc.org.uk>); *Mycologia* is the journal of the Mycological Society of America (<http://msafungi.org>); *Mycotaxon* is a journal dedicated to fungal taxonomy, and *Fungal Genetics and Biology* is an excellent source for experimental studies. Research on mycorrhizal fungi appears in *Mycorrhiza*, and work on plant pathogenic fungi is published in *Phytopathology*, *Plant Pathology*, and *Annual Reviews in Phytopathology*. Studies on fungi and human disease are published in *Medical Mycology* and other clinical journals. Reviews and essays on the most important contemporary topics in fungal biology are also published in *Microbiology and Molecular Biology Reviews*, *Microbiology Today*, *Nature Reviews Microbiology*, and *Trends in Microbiology*.

Textbooks

C. J. Alexopoulos, C. W. Mims, and M. M. Blackwell, *Introductory Mycology*, 4th edition (New York: Wiley, 1996).
S. C. Watkinson, N. P. Money, and L. Boddy, *The Fungi*, 3rd edition (Amsterdam: Elsevier, 2016).

J. Webster and R. W. S. Weber, *Introduction to Fungi*, 3rd edition (Cambridge: Cambridge University Press, 2007).

The following website is a useful supplement to these books because it presents a clickable evolutionary tree that directs readers to details on individual groups of fungi: <http://tolweb.org/tree/phylogeny.html>.

Books for non-specialists

Books written for non-specialists include E. Bone, *Mycophilia: Revelations from the Weird World of Mushrooms* (Emmaus, PA: Rodale Books, 2013), N. P. Money, *Mr. Bloomfield's Orchard: The Mysterious World of Mushrooms, Molds, and Mycologists* (Oxford: Oxford University Press, 2002), and N. P. Money, *Mushroom* (Oxford: Oxford University Press, 2011).

Amateur mycology societies

The North American Mycological Association (NAMA) is a vibrant society of mycological enthusiasts that organizes a popular annual foray (<http://www.namyco.org>). Local mycological societies publish newsletters that can be accessed online. An independent magazine, *Fungi*, is a highly informative quarterly journal on all things mycological (<http://www.fungimag.com>).

Mushroom identification guides

G. A. Lincoff, G. H. Lincoff, and C. Nehring, *National Audubon Field Guide to North American Mushrooms* (New York: Knopf, 1981).

K. McKnight and V. McKnight, *Peterson Field Guide Series: A Field Guide to Mushrooms of North America* (Boston: Houghton Mifflin, 1998).

J. Petersen, *The Kingdom of Fungi* (Princeton: Princeton University Press, 2012).

R. Phillips, *Mushrooms and Other Fungi of North America* (Buffalo, NY: Firefly Books, 2010).

P. Sterry and B. Hughes, *Collins Complete British Mushrooms and Toadstools: The Essential Photograph Guide to Britain's Fungi* (London: Collins, 2009).

These can be supplemented with regional guides and books on
particular groups of fungi. Online resources for mushroom
identification include:
<http://www.mushroomexpert.com>
<http://www.rogersmushrooms.com>

Index

Index

Fungi

Expand your collection of
VERY SHORT INTRODUCTIONS

SOCIAL MEDIA
Very Short Introduction

Join our community
www.oup.com/vsi

- Join us online at the official Very Short Introductions **Facebook** page.
- Access the thoughts and musings of our authors with our online **blog**.
- Sign up for our monthly **e-newsletter** to receive information on all new titles publishing that month.
- Browse the full range of Very Short Introductions online.
- Read **extracts** from the Introductions for free.
- Visit our library of **Reading Guides**. These guides, written by our expert authors will help you to question again, why you think what you think.
- If you are a teacher or lecturer you can order inspection copies quickly and simply via our website.